Global Environmental Change
An Atmospheric Perspective

John Horel and Jack Geisler
Department of Meteorology
University of Utah

John Wiley & Sons, Inc.
New York • Chichester • Brisbane • Toronto • Singapore • Weinheim

ACQUISITIONS EDITOR Nanette Kauffman
EXECUTIVE MARKETING MANAGER Catherine Faduska
SENIOR PRODUCTION EDITOR Elizabeth Swain
DESIGN A Good Thing Inc.
MANUFACTURING MANAGER Mark Cirillo
ILLUSTRATION COORDINATOR Edward Starr

This book was set in 10/12 Sabon by University Graphics, Inc. and
printed and bound by Quebecor. The cover was printed by Phoenix Color.

Recognizing the importance of preserving what has been written, it is a
policy of John Wiley & Sons, Inc. to have books of enduring value published
in the United States printed on acid-free paper, and we exert our best
efforts to that end.

The paper in this book was manufactured by a mill whose forest management programs include
sustained yield harvesting of its timberlands. Sustained yield harvesting principles ensure that
the number of trees cut each year does not exceed the amount of new growth.

Library of Congress Cataloging-in-Publication Data
Horel, John DeWitt.
    Global environmental change : an atmospheric perspective / John
Horel and Jack Geisler.
        p. cm.
    Includes index.
    ISBN 0-471-13073-7 (pbk. : alk. paper)
    1. Global environmental change. 2. Ozone layer depletion.
3. Global warming. 4. Internet (Computer network) I. Geisler,
Jack. II. Title.
GE149.H67 1997   96-32536
363.7—dc20   CIP

Printed in the United States of America

10 9 8 7 6 5 4 3 2 1

Preface

This book focuses on two aspects of global environmental change: global warming and the depletion of stratospheric ozone. It is intended as a text for a beginning-level undergraduate course in the broad curricular area of environmental sciences. Its emphasis is that of physical science applied at a very basic level to the atmosphere. A course based on the material presented in this text has been taught at the University of Utah during the past several years.

This book differs in several respects from others that include some treatment of global change at an elementary level. First, this is not a broad survey of the many ways that human activities are altering the global ecosystem. We confine our attention to global warming and to ozone depletion. What we specifically try to do is explain how the atmosphere works in this context of global change. The question addressed is this: What are the atmospheric processes that come into play to produce a global warming or to damage the protective layer of ozone that surrounds our planet? To answer this question requires the development of a foundation of knowledge of basic physical concepts related to the structure and behavior of the atmosphere and ocean. This foundation includes discussion of the natural climate variations associated with atmosphere–ocean interactions and volcanic eruptions as well as the skill of predicting weather and climate changes. Finally, a novel aspect of this text is its Internet Companion, an up-to-date supplement of information on the Internet created specifically to accompany this book. It provides access to thousands of pages of additional information.

Global warming is an anticipated consequence of present and future emission of certain types of greenhouse gases into the atmosphere as a result of human activities. One can put forward the explanation that the accumulation of these specific gases "creates a blanket around the earth" and that's why warming follows. What actually happens in the atmosphere in response to the accumulation of these gases is, to the surprise of no one, much more complex than that. Why should anyone other than a career scientist try to understand how this response actually works? In our opinion, this is the only way to appreciate why global warming is such a controversial issue. Some say it is happening now, some say it is not, and some say that it isn't going to happen. What is the basis for this diversity of opinion?

Depletion of stratospheric ozone is the present consequence of emission into the atmosphere of a gas that was specifically designed in a laboratory and subsequently produced for many different industrial applications. The chain of events in the atmosphere that leads from this emission to ozone depletion is much better understood than is the chain of events that lead from emission of greenhouse gases to global warming. Indeed, the understanding is so complete as to underwrite international agreement that will phase out commercial production of this type of gas in many countries as of the end of 1995. In this country at this stage of our economic evolution, the outright ban by government of an industrial product on the grounds of protecting something as intangible and remote from experience as ozone in the stratosphere is remarkable. The opinion that the ozone-depletion problem is overrated or even a hoax is increasingly heard. As with global warming, we feel that an opinion on the ozone-depletion problem requires some understanding of what goes on in the atmosphere to produce it.

THE INTERNET COMPANION

The methods for exchange of information are developing rapidly. The expanding use of the Internet for dissemination of information is particularly apparent in the atmospheric sciences. Up-to-the-minute weather and climate information can be received on the instructor's desk-top computer, displayed directly on wide-screen projection devices in the classroom, or viewed by the student in a computer laboratory or on a home computer. The subjects of global warming and ozone depletion are uniquely suited to exploit this developing technology. Because they are subjects of considerable interest to scientists, policy makers, and the general public, there is a vast amount of information available about them. Government agencies provide daily images of the ozone hole over Antarctica, while environmental organizations provide forums for discussions of policy issues.

The availability of this information via the Internet simplifies the requirements for graphics and information presented in the text. The black-and-white figures in the text are comparatively few relative to the total number of figures usually presented in introductory texts, many of which are often costly color images. Now the instructor or student can obtain hundreds of relevant figures on their own in color over the Internet. Plus, other media forms, such as video clips, can be accessed and integrated into presentations.

Now that we have explained why the Internet Companion is available as a supplement to the text, it is relevant to ask why the text itself is necessary, if so much information is available already via the Internet. Searching for information on the Internet can be an absorbing and daunting task. Many analogies exist for the present state of the Internet. One of our favorites is to view the sources of information on it as every book in a major university library thrown in a heap onto the university's football field. How would you find the information that you want and how do you place the information that you find in context? While there are many ways to help organize information on the Internet, this text and the accompanying Internet Companion are means to explore the subjects of global warming and ozone depletion in a constructive and stimulating manner for both instructors and students.

The Internet Companion is part of a World Wide Web server located in the Department of Meteorology at the University of Utah. For those already familiar with the jargon used on the Web, the Internet Companion is located at the following Uniform Resource Locator (URL): http://www.met.utah.edu/climate.html. If this terminology is unfamiliar to you, it is explained further in Appendix A.5. The Internet Companion consists of several types of information:

1. up-to-date color copies of figures used in the text;
2. other figures that expand directly on the material presented in the book, but, because of space limitations, could not be included in the text;
3. access to large amounts of information provided by government agencies and scientists; and
4. access to forums for discussions of opinions on climate change issues.

The Internet Companion can be used independently to browse for information related to climate change. However, to be able to place the information in context requires relying on the accompanying text. At the end of each chapter of the book, suggested locations in the Internet Companion for additional information are presented. This replaces the bibliography section frequently used in introductory texts. The locations in some cases may point to a single figure, but most often will contain interactive access to in-depth information.

The user of the Internet Companion must become familiar with the restrictions on the access, use, and dissemination of electronic information subject to copyright restrictions. For example, the color copies of the figures used in this text are accessible to everyone for personal use, in the same manner that anyone can look at a book at the library. However, they cannot be reproduced or distributed without the permission of the publisher. Most sites on the Internet place similar restrictions on their information.

ACKNOWLEDGMENTS

We would like to thank Leslie Allaire for her patience, diligence, and good humor while typing several drafts of this text, obtaining copyright approvals, and handling many other tasks related to this work. Bryan White played a major role in the design and development of the Internet Companion to this text. Steven Krueger, Jan Paegle, and Julia Nogués Paegle provided many valuable comments based on their classroom experiences that relied on earlier drafts of the text.

We would also like to thank the following reviewers for their valuable comments and suggestions: Thomas J. Crowley, Texas A&M University; Julie Elbert, University of Southern Mississippi; David S. Gutzler, University of New Mexico; John Harrington, Jr., Kansas State University; Daniel J. Leathers, University of Delaware; Jeffrey Rogers, Ohio State University; Daniel Wise, Western Illinois University.

The scientists involved in the Working Groups of the Intergovernmental Panel on Climate Change are to be commended for their efforts in defining and describing the scientific issues associated with global warming. We relied heavily on their work.

Contents

CHAPTER 1 **The Global Atmosphere** 1
The Elements of Global Warming 2
An Example of Climate 4
The Global Surface Temperature Record 6
Review Questions 9
Internet Companion 9

CHAPTER 2 **The Radiation Balance of the Earth-Atmosphere System** 11
Calculating the Radiation Balance 12
Simple Models of the Greenhouse Effect 16
The Effect of Clouds on the Radiation Balance 21
Review Questions 24
Internet Companion 25

CHAPTER 3 **Weather and Climate** 27
The Weather 27
Satellite Cloud Imagery 32
The Prediction of Weather and Climate 36
The Hydrologic Cycle 45
Review Questions 47
Internet Companion 47

CHAPTER 4 **The Natural Variability of the Earth-Atmosphere System** 49
The Role of Volcanoes in the Earth-Atmosphere System 49
Air-Sea Interactions 55
Review Questions 65
Internet Companion 65

CHAPTER 5 **Long-Term Climate Variations** 67
Paleoclimatology 67
The Ice Ages 72
The Milankovitch Theory of the Ice Ages 75
Review Questions 81
Internet Companion 81

CHAPTER 6   The Carbon Cycle   83
The Preindustrial Carbon Cycle   83
The Geochemical Carbon Cycle   90
The Present Carbon Cycle   91
Review Questions   96
The Internet Companion   96

CHAPTER 7   Greenhouse Gases, Clouds, and the Radiation Balance   97
The Other Greenhouse Gases   97
Radiative Forcing Due to Increase of Greenhouse Gases   100
Aerosols and Sulfur   103
Radiative Forcing by Clouds   106
Review Questions   110
Internet Companion   111

CHAPTER 8   Predicting Climate Change   113
A Review   113
The Feedback Processes   114
Doubling Carbon Dioxide in Climate Simulations   116
Risk Versus Uncertainty   123
Review Questions   126
Internet Companion   126

CHAPTER 9   Depletion of Stratospheric Ozone   127
The Natural Life Cycle of Ozone   128
The Antarctic Ozone Hole   131
Ozone Depletion as a Global Environmental Issue   136
The Future   139
Review Questions   141
Internet Companion   141

APPENDIX   143
About Numbers   143
About Units   143
About Graphs   144
Some Basic Chemistry   144
Access to the Internet Companion   147

INDEX   149

CHAPTER 1

The Global Atmosphere

A familiar aspect of the atmosphere in which we live is that cooler climates are found at higher altitudes. This fact is contained in the following formal statement: The temperature decreases with altitude through a layer of the atmosphere extending from the earth's surface up to an average altitude of about 10 km (33,000 ft). This layer of the atmosphere is called the *troposphere*. It contains about 80 percent of the air that makes up the atmosphere and virtually all of its weather and its clouds. Immediately above the troposphere lies the layer of the atmosphere known as the *stratosphere*.

The division of the atmosphere into troposphere and stratosphere is also a fundamental one in the organization of this book. Just about 80 percent of it is devoted to things that are going on in the troposphere. This is where the mechanism of global warming is located. The component parts of that mechanism are many, and we have to look at each one separately before we can get some feeling for how they work together. The troposphere is also that part of the atmosphere in contact with the earth's surface. People have recorded its influence at the earth's surface over recent centuries; the impact before that is indicated in the geologic record. Global warming is certainly global climate change, and assessing change involves defining a standard against which the change is determined. For this standard we have to look to the record of past global climate conditions. Something about this record, therefore, has to appear in a book that treats global warming.

Thus it happens that consideration of the stratosphere, where the problem of ozone depletion is occurring, is relegated to only one chapter of the book. Such a situation is contrary to the importance and urgency of this global environmental problem, which has been clearly shown to be a direct consequence of human activities. On the other hand, the relative simplicity of the mechanism of stratospheric ozone depletion and the advanced state of scientific understanding and description of it both make it possible to tell the tale with a brevity not possible in the case of global warming.

1.1 THE ELEMENTS OF GLOBAL WARMING

The earth as seen from space by the eye or in a color photograph shows white clouds set against a background of blue oceans and the occasional continental land mass. All of this is seen as a result of visible light from the sun that is *reflected* back out to space. The visible light that is not reflected back to space is absorbed in the atmosphere and at the surface of the earth. Although visible light constitutes a large part of the spectrum of electromagnetic radiation that is emitted by the sun, the solar beam also includes some radiation in the ultraviolet and in the infrared parts of the spectrum. This radiation, which is not visible to the eye, is absorbed by the atmosphere.

Averaged over the entire globe and over the period of a year, the reflected visible light amounts to about 30 percent of the solar radiation that is intercepted by the earth and the atmosphere. The remaining 70 percent that is absorbed constitutes a source of heat for the earth and the atmosphere. Yet, we know from observations that global temperature remains remarkably constant from year to year. Specifically, the temperature of the air at the earth's surface, which we call in this book as the *surface temperature*, when averaged over the globe and over the year, has varied less than 1°C (about 2°F) over the past century. What maintains this equilibrium is that the earth and the atmosphere are *emitting* radiation out to space at a rate that balances the 70 percent of intercepted solar radiation that is being absorbed. This emitted radiation is entirely in the infrared part of the electromagnetic spectrum.

The rate at which infrared radiation is emitted out to space depends on the composition of the atmosphere. Of particular importance are the so-called *greenhouse gases*. The direct result of an increase in the amount of a greenhouse gas in the atmosphere is a decrease in the rate of emission of infrared radiation from the atmosphere out to space. There is then a temporary imbalance between solar radiation absorbed globally and infrared radiation emitted globally to space. Other things being equal, this temporary imbalance is removed by a global increase of atmospheric temperature. This is *global warming*.

Carbon dioxide is a greenhouse gas. The amount of it in the atmosphere has risen by about 30 percent due to human activities since the beginning of the Industrial Revolution. Its continuing increase in the global atmosphere is illustrated in Figure 1.1. The present concentration of 360 parts per million (ppm) means that 360 of every million molecules of gas in the atmosphere are carbon dioxide. Projected world economic growth is such that the present amount of carbon di-

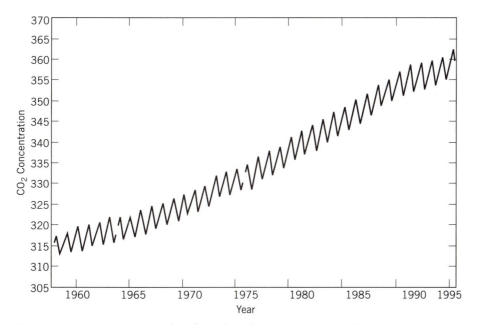

Figure 1.1 Concentration of carbon dioxide at Mauna Loa, Hawaii, in parts per million (ppm).

oxide in the atmosphere will double (increase by 100 percent) before the end of the next century unless specific measures are taken to limit its generation.

Why is carbon dioxide increasing? Increased industrialization of the planet-wide economy is the predominant cause. Burning fossil fuels (oil, gas, coal, etc.) is one way that carbon dioxide is released into the atmosphere, while another source is the burning of timber and vegetation. Trees and plants remove carbon dioxide from the atmosphere in the course of the photosynthesis process. Note the seasonal changes in the concentration of carbon dioxide in Figure 1.1. The amount of vegetation is larger in the Northern Hemisphere than in the Southern Hemisphere (oceans occupy a smaller fraction of the Northern Hemisphere than the Southern Hemisphere). What the seasonal oscillation seen in Figure 1.1 represents is the removal of carbon dioxide from the atmosphere in Northern Hemisphere spring and summer and its return in autumn and winter.

Other greenhouse gases in the atmosphere are also increasing as a result of human activities. What global warming has already occurred as a consequence of greenhouse-gas increase? What global warming is projected to occur in the coming century? We will look at the answers toward the end of this book. These answers are tentative and subject to uncertainty; that is, they are estimates. A more immediate goal of this book is to provide a basis for interpreting these estimates. We stated that global warming (in some amount) is the response to an increase of the amount of a greenhouse gas in the atmosphere, other things being equal (that is, remaining the same). What are these other things, and do they in fact remain the same while greenhouse gases increase?

The atmosphere is, in many respects, a random system. In other words, any particular spatial or temporal scale is as likely as any other. For example, storms pass across the United States at irregular intervals. There are periods during which storms will pass across the country every few days; this may be followed by a period during which storms are separated by a week or more. On the other hand, the atmosphere is *forced* at specific temporal and spatial scales. "Forced" refers to driving by external physical processes. Thus, the atmosphere as a system is forced by changes in the characteristics of the underlying land and ocean surfaces. The only significant natural processes that are external to the combined earth-atmosphere system as a whole have to do with the sun. The earth-atmosphere system is forced primarily at specific temporal and spatial scales by changes in the amount of sunlight incident on the earth's surface, as will be discussed in greater detail later.

Stretching our terminology a bit, we can refer to human-induced forcing of the earth-atmosphere system as unnatural, in the sense that the forcing would not occur without the intervention of people. Because using the term "unnatural" for human activity is a little odd, we will follow convention and refer to it as anthropogenic forcing. Some forms of anthropogenic forcing require little technology, for example, slash and burn agriculture or overgrazing. Others result from the increasing global dependence on the burning of fossil fuels and other industrial by-products.

The combination of random and forced behavior leads to chaotic behavior in the earth-atmosphere system. Edward Lorenz, an atmospheric scientist of considerable renown, first investigated the chaos phenomenon that has now been studied in many other scientific and socioeconomic fields. The characteristic of chaos that is both compelling and frustrating is the elusive sense of order in the midst of random behavior. As we will see in Chapter 3, it is the chaotic nature of the atmosphere that limits our ability to predict the weather more than a few days in advance and contributes to uncertainty regarding our confidence in the prediction of climate many years into the future.

1.2 AN EXAMPLE OF CLIMATE

We used the term "global warming" in the preceding section to refer to a global increase in the temperature of the atmosphere. In this book we will follow the common practice by adopting as the measure of any such temperature increase in the atmosphere as a whole the increase of air temperature at the surface. We are thus defining global warming as an increase in the global-average surface temperature. The global-average surface temperature then takes on the role of a measure, or index, of global *climate*. But climate, whether it be local or global, is a time-average concept, and we must specify the length of the averaging time when we talk about it. If we talk about climate change (in our case, a change in global-average surface temperature), then two averaging times have to be disclosed. One is the time period for which the climate is defined and the other is the (longer) time period that defines what is the normal or baseline climate against which we determine whether the climate has changed.

We illustrate these two time averages by consideration of a local climate defined by surface temperature at a weather station. On a given night on the local news, a weathercaster states that, for example, the "normal" high for today at Salt Lake City, Utah, is 45°F and that the normal low is 25°F. What is meant by "normal" here? It means the expected or the usual temperature. This expected high is officially determined by averaging over the 30-year period 1961–1990 the high values that occurred on that day of the year. Similarly, the expected low is the 30-year average of the low values occurring on that date.

The 30-year average that defines "normal" in this example is the longer of the two time averages. Consider now the shorter of the two time averages. This defines the climate whose change, if any, we wish to determine. For purposes of this example, we take this to be the annual average. First we note that the daily average temperature at a station is defined to be the average of the observed high and the observed low. The normal daily average temperature for the example of a high of 45°F and a low of 25°F is 35°F. If we take a one-year record of these daily averages and average them over the year, we have the annual-average temperature for that year. Figure 1.2 shows this at Salt Lake City over the period from 1930 through 1995. Note that it has varied within the rather narrow range of 48°F to 56°F. There is no readily discernible warming or cooling trend over this period, hence no evidence of climate change.

The normal annual-average temperature is the average over the 30-year period 1961–1990 of the annual averages from Figure 1.2. This turns out to be 51.8°F. If we subtract this normal from the annual average for any particular year, we have what is known as the *anomaly* for that year. We can readily convert the data in Figure 1.2 into anomalies. The result is shown in Figure 1.3. The display in terms of anomalies renders a little clearer the absence of any warming or cooling trend in Figure 1.2.

When we take the annual-average temperature at reporting stations such as Salt Lake City and average these numbers over the entire globe, we have the annual- and global-average surface temperature. This, as we noted in the preceding section, has varied by less than 2°F over the past century. However, this rather small variation does exhibit a warming trend. We will look at this trend in the next section.

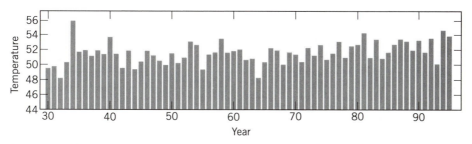

Figure 1.2 Annually averaged surface temperature (in °F) at Salt Lake City from 1930 through 1995.

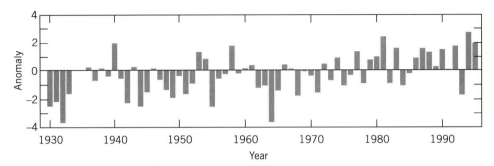

Figure 1.3 Departures of annually averaged surface temperature (in °F) at Salt Lake City from the 1961–1990 average.

1.3   THE GLOBAL SURFACE TEMPERATURE RECORD

The temperature of the air near the earth's surface, which we have agreed to call simply the surface temperature, has been routinely monitored at many points over the globe for more than a century. The data have come from ships at sea as well as from observing stations on land. In the 1940s, these surface temperature data began to be supplemented with temperature data from balloons released twice daily at weather stations to monitor temperatures at other levels in the atmosphere. More recently these data have in turn been supplemented by temperatures at many different levels in the atmosphere remotely sensed by satellites. When these data are averaged over the globe and over several years, we have the temperature profile seen in Figure 1.4. The temperature decreases steadily from a surface value of 288 K (15°C, 59°F) to attain a minimum of 223 K at an altitude of about 10 km. As we said at the beginning of the chapter, this layer of the atmosphere through which the temperature decreases with altitude is called the troposphere. Above 10 km the temperature increases with altitude and continues to do so up to a level of about 50 km. This layer of the atmosphere is called the stratosphere.

We have said that the global-average surface temperature is what is used as a measure of global warming in preference to, say, the global-average temperature from Figure 1.4 further averaged over the depth of the troposphere. This would be more representative of the temperature of the troposphere as a whole than the surface temperature alone would be. Two factors preclude reliance on this more representative set of temperature data. One is that the temperature measurements through the depth of the troposphere have been limited by available technology and are less accurate than those measured by surface-based thermometers. The other is that the length of the record is considerably shorter than the length of the surface-temperature record.

The surface-temperature record is shown in Figure 1.5. Each vertical bar represents the anomaly (in °C) in the global- and annual-average surface temperature relative to the 30-year normal, which in this case is the average over the period 1961–1990. This normal has the value 288 K (15°C, 59°F). The smooth curve is a "running mean" that irons out sharp variations from year to year. On top of

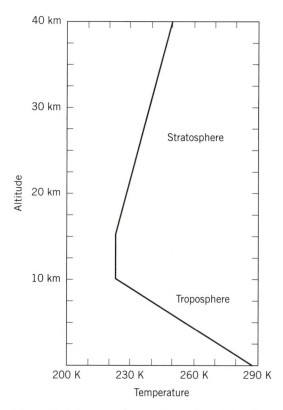

Figure 1.4 Averaged over the entire globe, the temperature decreases from 288 K to 223 K at the top of the troposphere and then gradually increases through the stratosphere.

all of this we have drawn a straight line to bring out the trend that characterizes the entire 135-year record.

The trend as given by our straight line is 0.55°C over the entire 135 years of the record. However, the smooth curve is a far better representation of this record than is the straight line. It shows, first of all, that the global-average temperature exhibited very little trend from 1860 to the very early 1900s. If we follow the smooth curve from there, it indicates a pronounced upward trend from about 1910 through 1940. The next few years saw a small decrease followed by a period lasting until about 1980 in which there was no trend. The rise in global average temperatures characteristic of the period 1910 through 1940 then returned and remained throughout the decade of the 1980s that led to a peak temperature in 1990. Then the global average temperature dropped over the next two years before reaching a new record temperature in 1995.

There is some controversy as to whether the record shown in Figure 1.5 constitutes evidence of global warming. We could try to convince ourselves that it does by redrawing the figure with the time axis compressed. Such a figure is shown in the appendix as Figure A.2. The sharp rise in the first part of the century, which

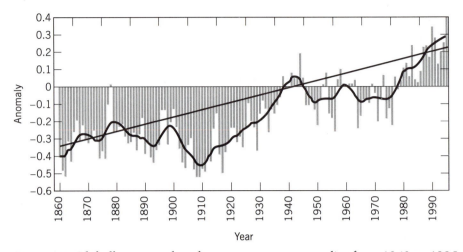

Figure 1.5 Globally averaged surface temperature anomalies from 1860 to 1995 (in °C) compared to the 1961–1990 average. The heavy line smooths out the year-to-year variations in the record while the straight line reflects the 135-year trend.

resumes in the 1980s, would then look a lot more threatening. Even so, the record definitely lacks the quality of inexorable rise that has characterized the amount of carbon dioxide in the atmosphere (Fig. 1.1). Global warming in response to rising carbon dioxide, if any, has occurred in a most irregular fashion.

The surface-temperature record itself is subject to uncertainty. To begin with, a number of reporting stations have moved to new locations over the years. Others have found themselves enveloped by urban development—a serious concern, because cities are notably warmer than the surrounding countryside. About 70 percent of the surface of the globe is ocean. A lot of the data in the over-ocean part of the record consists of surface air temperatures inferred from measurements of ocean temperature. Moreover, in the time span of the record, ocean shipping has gone from sail to power, and this has changed the data coverage. Ships today use rather narrow shipping lanes that are defined as the shortest distance between two ports. Sailing vessels gave a much better spatial coverage of the global ocean.

Much effort has gone into correcting the global-average temperature record to minimize the errors introduced by these and other sources. For better or worse, we accept it for what it is. What does it tell us? There is a consensus among scientists that this record shows that there has been an increase in global-average surface temperature in the amount of roughly 0.5°C over a 100-year period. There is also a widely shared opinion that this is not a large increase. The preponderance of evidence now suggests, however, that this trend in surface temperature is unlikely to be entirely natural in origin. In other words, human activities have affected the global climate.

We must emphasize that this conclusion is one of consensus. It represents the view of many, but by no means all, in a large group of scientists working across a broad range of disciplines. Given the nature of consensus, it is perhaps not surprising that there has not been any action to limit emissions of carbon dioxide. Because it is relatively more abundant in the atmosphere than other greenhouse gases generated by human activities, carbon dioxide has the most potential for producing global warming. Its rise continues.

How much carbon dioxide can the atmosphere hold before we do get un-equivocal evidence from the surface-temperature record that global warming has occurred? To answer this question we have to rely on calculations done in the framework of models. Models incorporate mechanisms in the atmosphere that convert greenhouse gas increase into global warming. In the next chapter we ex-amine the basic atmospheric mechanism that, when perturbed by an increase in greenhouse gas, yields global warming.

R E V I E W   Q U E S T I O N S

1. What distinguishes the troposphere from the stratosphere?
2. Why should it be expected that the seasonal variation in CO_2 is less at the South Pole than in the middle-latitudes of the Northern Hemisphere? Verify your answer by using the suggested information in the Internet Companion.
3. What is the global- and annual-average surface temperature at present?
4. How has the global- and annual-average surface temperature changed during the past 135 years?

INTERNET COMPANION

- Refer to Appendix A.5 for information on how to access the Internet Companion. Here are some locations of particular interest.

1.1   The Elements of Global Warming

- The Carbon Cycle Group of the Climate Monitoring and Diagnostics Laboratory of the National Oceanographic and Atmospheric Administration (NOAA) provides data on concentrations of CO_2 at four sites since 1974. This information can be used to see how the increase over time in CO_2 concentration is fairly uniform around the globe when averaged over an entire year.

1.2   An Example of Climate

- Further information on the weather and climate of Salt Lake City and other locations is made available by a number of agencies.

1.3   The Global-Surface Temperature Record

- Current information on the trend of the surface temperature averaged over the globe is provided from the World Meteorological Organization and other government organizations.

1.4   Other Interesting Web Sites

- The NOAA Office of Global Change Programs provides an overview of research underway to study the issues of global warming.

- An on-line magazine, *Consequences: The Magazine,* discusses current issues related to climate change. Greenpeace provides one perspective on the global warming problem.

- The consensus view on global warming of the scientific community can be found in the reports of the Intergovernmental Panel on Climate Change.

CHAPTER 2

The Radiation Balance of the Earth-Atmosphere System

A *radiation balance* exists when the radiation that the earth (together with its atmosphere) emits to space is equal to the solar radiation that it absorbed. No such balance exists from day to day or even from season to season, but over a period of a full year, the balance used to be very nearly perfect. It may still be. There is every reason to suspect that it is not, due to the increased amounts of greenhouse gases in the atmosphere that are the collective result of human activities since the advent of the Industrial Revolution. As world economic development continues, the rate at which greenhouse gases are accumulating in the atmosphere is itself increasing. In this inflationary scenario, near-perfect radiation balance is sure to end, if it has not done so already.

Global warming is a response to the existence of the type of radiation *imbalance* that a greenhouse-gas increase is capable of producing. In this chapter we will see what type of imbalance that is. Before we look at an imbalance, we have to consider the state of perfect balance. In particular, we need to set down some numbers that characterize the balance. These established, we go on to develop a simple model that explains the basic atmospheric mechanism that results in global warming. We conclude the chapter with a section on how clouds affect the radiation balance.

2.1   CALCULATING THE RADIATION BALANCE

The first thing that we have to do is to arrive at a numerical value for the solar radiation that the earth receives from the sun. Consider first an easily visualized analogy. A 100 watt lightbulb is enclosed by a spherical lamp shade with a radius (distance from the center of the bulb to the lamp shade) of 0.5 meters, (or about 20 inches). This is illustrated in Figure 2.1. (Standard abbreviations of measurements such as the watt [W] and the meter [m] are listed in Appendix A.2.)

Intensity

Intensity is defined as the radiation received by an object divided by the area of the object. In this example, the object will be the inside of the lamp shade. The power of the light source is 100 W, and that is also the power received by the entire inside of the lamp shade. The area of the lamp shade can be expressed mathematically as $4\pi R^2$ (the area of a sphere with radius R), and R equals 0.5 m here. The intensity of light received by the inside of the lamp shade is thus

$$\frac{100 \text{ W}}{4 \times 3.14 \times (0.5 \text{ m})^2} = \frac{100 \text{ W}}{3.14 \text{ m}^2} = 32 \text{ W m}^{-2}$$

Now imagine that the inside of the lamp shade is painted black and therefore absorbs all the light that shines on it. The result of the above equation is that 32 W would be absorbed for every square meter of lamp-shade surface. How much power would be absorbed by a tiny circular patch shown by the black spot (Fig.

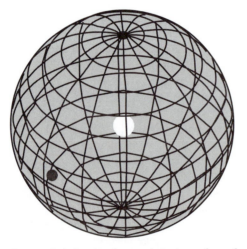

Figure 2.1 A wire frame drawing of a spherical lamp shade that surrounds a light (here represented by the shaded orb in the middle). The circular patch in the lower left corner occupies a hundredth of the surface area of the lamp shade.

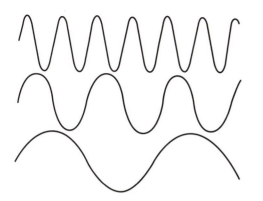

Figure 2.2 The wavelength of a wave is measured from crest to crest or trough to trough. In this example, the top, middle, and bottom waves can be characterized as short, medium, and long waves, respectively.

2.1) on the shade and having an area that is a hundredth of the area (3.14 m²) of the shade? The patch would absorb 1 W of power.

The power emitted by the bulb spans a portion of the electromagnetic radiation spectrum. Specific sections of this spectrum are referred to in terms of the wavelength of the radiation. All radiation travels in a fashion that can be thought of as a wave. Imagine for a moment that the radiation is traveling toward you. Every so often, a wave "crest" will pass you, followed by a wave "trough" as shown in Figure 2.2. The length of the wave (or wavelength) can be defined as the distance between successive crests or troughs. For our purposes, a useful measure of the wavelength of radiation is a micron (μm), which is a millionth of a meter. Most of the radiation emitted by the lamp is *visible* light, which has wavelengths between 0.4 μm (violet) and 0.7 μm (red). A section of the spectrum of electromagnetic radiation is shown in Figure 2.3. Some of the radiation emitted by the lamp lies at shorter wavelengths than what can be detected by the human eye. This type of radiation is referred to as *ultraviolet* radiation. Some radiation is also emitted at longer wavelengths than visible light, and this is referred to as *infrared* radiation.

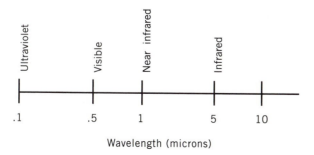

Figure 2.3 Portions of the electromagnetic spectrum are distinguished by their wavelength.

The Absorbed Radiation

The power output of the sun is 3.87×10^{26} W, a large number written here in the exponential notation that is explained in Appendix A.1. We can use the example of the lightbulb and the tiny patch on the lamp shade to determine how much power is available to the earth. Consider again Figure 2.1, where the sun now functions as the light bulb. The lamp shade is replaced by an imaginary sphere whose radius R is the distance from the center of the earth to the center of the sun. This distance is 1.5×10^{11} m. The intensity of the solar radiation received on the imaginary sphere is

$$\frac{3.87 \times 10^{26} \text{ W}}{4 \times 3.14 \times (1.5 \times 10^{11} \text{ m})^2} = 1368 \text{ W m}^{-2}$$

This number is known as the *solar constant*.

The earth as seen from the sun appears as a flat disc pasted onto the imaginary sphere, just as the sun and moon appear to us as flat discs pasted onto the sky. From the standpoint of the radiation that falls on it, this earth-disc is equivalent to the patch stuck on the inside of the lamp shade. The power received by the earth from the sun is then the product of the intensity of the solar radiation (1368 W m^{-2}) and the area of the earth-disc (πr^2, where r, the radius of the earth, is 6.37×10^{6}m). However, we must also consider that, since the earth rotates around its axis every 24 hours, the power received is spread out over a spherical surface. We thus divide the received power by the area of the spherical surface of the earth to arrive at the intensity of the solar radiation received per unit area of the earth's surface

$$\frac{1368 \times 3.14 \times (6.37 \times 10^6 \text{ m})^2}{4 \times 3.14 \times (6.37 \times 10^6 \text{ m})^2} = \frac{1368}{4} = 342 \text{ W m}^{-2}$$

A major difference between the patch on the lamp shade and the earth is that we specified that the lamp shade was black; that is, all the incident radiation received by the patch was absorbed. However, some of the visible solar radiation received by the earth-atmosphere system is reflected back out to space, giving the blue-white image of the earth as seen from space. The amount reflected back out to space turns out to be 30.7 percent of the total incident radiation. This fraction of the total radiation that is reflected out to space is referred to as the earth's *albedo*. In terms of the intensity of radiation, 105 W m^{-2} (0.307×342 W m^{-2}) is reflected out to space and the remaining 237 W m^{-2} (0.693×342 W m^{-2}) of the incident radiation is absorbed. This is summarized in Figure 2.4.

The Emitted Radiation

We have just calculated that, when averaged over the entire surface of the earth, the intensity of absorbed solar power is 237 W m^{-2}. In this calculation we used the annual-average distance of the earth from the sun ($R = 1.5 \times 10^{11}$ m). Thus, to be quite specific, the number 237 W m^{-2} is the intensity of absorbed solar power *in the global and annual average*.

At this point it is necessary to start thinking in terms of the *earth-atmosphere*

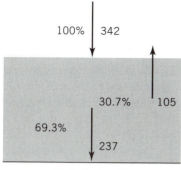

100%   342

30.7%   105

69.3%

237

Surface of earth

Figure 2.4 Of the incoming solar radiation 69.3% (237 W m^{-2}) is absorbed by the earth-atmosphere system while 30.7% (105 W m^{-2}) is reflected back out to space.

system. This is composed of the atmosphere, the underlying solid surface (land and ice), and the entire global ocean. In the global and annual average, the intensity of the solar power reflected by the earth-atmosphere system is, as we have seen, 105 W m^{-2}. Most of this reflection actually occurs in the atmosphere, and clouds are the agent that accomplishes it. The solar power that is not reflected, whose intensity in the global and annual average is 237 W m^{-2}, is absorbed by the earth-atmosphere system. A small fraction is absorbed in the atmosphere, but most of it reaches the earth's surface and is absorbed there. In the interest of simplification, we will ignore the relatively small fraction absorbed in the atmosphere and simply assume that all solar radiation that is not reflected back to space is absorbed at the earth's surface.

Let us assume for the time being that there is no global warming or cooling of the earth-atmosphere system in the annual average (that is, over the course of the year). Then the 237 W m^{-2} of incoming solar power absorbed must be balanced by 237 W m^{-2} emitted to space by the earth-atmosphere system. We say that the system is in *radiative balance.* There is an important difference in the respective two kinds of radiation that are involved here. The solar radiation comes from a very hot source and is mostly in the visible part of the electromagnetic spectrum. The earth-atmosphere system has a much lower temperature than the sun, and as a result, the radiation that it emits to space is entirely in the infrared part of the electromagnetic spectrum.

Infrared radiation cannot be seen by the eye. It can, however, be felt as heat. An electric-range burner unit at a low setting has no glow, but it emits power in the infrared. You can feel it if you place your hand a short distance away from the burner unit. At a high setting, the unit will emit some red (visible) light, but most of the radiation is still in the infrared. From the standpoint of power consumption, the electric range is a device that converts electric power into infrared radiation. From the standpoint of solar power consumption, the earth-atmosphere system is a device that converts radiation in the visible part of the spectrum into radiation in the infrared part of the spectrum.

What temperature must the earth-atmosphere system have in order for the incoming absorbed solar radiation to be balanced by the outgoing infrared radiation? A fundamental law of physics, the Stefan-Boltzmann Law, states that the intensity of radiation emitted by an object is proportional to the fourth power of the temperature (in degrees Kelvin) of the radiating object. The constant of proportionality in this simple mathematical law has the numerical value of 5.67×10^{-8}. As is readily verified with a hand calculator, the temperature that an object must have to emit radiation with an intensity of 237 W m^{-2} is 254 degrees Kelvin (K).

We will refer to the temperature just calculated as the *equilibrium temperature* of the earth-atmosphere system. The system is radiating as if it were at a temperature of 254 K ($-19°C$, $-2°F$). This is certainly colder than the surface temperature that we looked at in Chapter 1. As we saw there, the observed global- and annual-average surface temperature is 288 K ($59°F$).

How do we reconcile the 34 K difference between the two temperatures 254 K and 288 K? As shown in Figure 1.4, the air temperature decreases away from the earth's surface at a rate equal to 6.5 K per km of elevation. So, the conclusion is that the earth-atmosphere system must radiate energy to space as if from a surface located somewhere up in the atmosphere. We will refer to the place where this surface is located as the *average effective radiating level*, which will be abbreviated to simply *average ERL*. To determine how high above the ground the average ERL is located, we divide 34 K by 6.5 K per km, which yields 5.2 km. This is about halfway up the troposphere (Figure 1.4). The reason that the average ERL is located in the atmosphere and not at the surface is that the atmosphere is opaque to infrared radiation. This means that the infrared radiation emitted by the surface (land or ocean) is absorbed by the atmosphere, and the infrared radiation that goes out to space is emitted by the atmosphere.

2.2. SIMPLE MODELS OF THE GREENHOUSE EFFECT

An Isothermal Model

We are now at a point where we can begin to use some of the concepts we have developed to understand how global warming can occur. We will simplify the situation to begin with by assuming that the layer of gas that comprises the atmosphere has the following properties:

- It contains no clouds and is completely transparent to radiation coming from the sun. The solar radiation that passes through it is absorbed at the earth's surface.
- It is completely opaque to infrared radiation emitted from the earth's surface. Such radiation must first be absorbed by the atmosphere and then re-emitted to space.
- Its temperature is the same everywhere (an *isothermal* atmosphere). We ignore the fact that the temperature in the real atmosphere decreases with height.

Figure 2.5 The radiation balance of an earth climate system with no atmosphere.

All of these conditions help to define a particular abstract view, or *model,* of the atmosphere. We will find this model useful in developing the concepts necessary for understanding the global warming problem.

We agreed in Chapter 1 to use the term "surface temperature" to refer to the temperature *of the atmosphere* at the surface of the earth. This is not, in general, exactly the same as the temperature of the underlying land (or sea) surface. In the model that we are presenting here we shall refer to the surface of the earth as the ground and to its temperature as the *ground temperature.* Finally, we impose on the model the restriction that there shall be a steady state, by which we mean that neither the temperature of the atmosphere nor the ground temperature is changing with time.

We will compare what occurs in this model with what would occur on an earth that doesn't have an atmosphere. In the case of no atmosphere (Figure 2.5), solar radiation (a) incident on the ground (heavy arrow) is balanced by infrared radiation (b) emitted to space (light arrow). The length of the arrows indicates the intensity or, equivalently, the amount of radiation emitted or absorbed.

Because of the imposed restriction of a steady state, the amount of infrared radiation emitted to space must balance (equal) the amount of solar radiation absorbed by the ground. As we have discussed, we can use the Stefan-Boltzmann Law to determine the ground temperature that is required for this balance. We will assume that the intensity of solar radiation absorbed is 342 W m^{-2}. As can be verified by hand calculator, the Stefan-Boltzmann Law then tells us that the ground temperature has to be 276 K (3°C, 37°F). This temperature is higher than the 254 K temperature that we determined earlier because we are not taking into consideration in this model the reflection of solar radiation that reduced the intensity of solar radiation absorbed to 237 W m^{-2}.

The model that contains both earth and atmosphere is illustrated in Figure 2.6. Because the situation is a steady state, the infrared radiation (f) emitted from the top of the atmosphere must equal the solar radiation incident at the top of the atmosphere (c). In addition, because the atmosphere is completely transparent to it, the solar radiation at the ground is the same as the solar radiation at the top of the atmosphere. This, in turn, is the same as the case in which there is no atmosphere (the arrow at (a) in Figure 2.5). However, the intensity of the total radiation (solar and infrared) absorbed by the ground is the sum of (c) and (e). This is obviously greater than the radiation (a) absorbed by the ground in the case of an earth with no atmosphere.

It follows from the imposed steady state that the radiation emitted by the

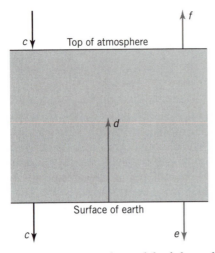

Figure 2.6 A simple model of the radiation balance of the earth-atmosphere system.

ground (d) must be equal to the sum of (c) and (e). Since the intensity emitted by a body depends on its temperature only (from the Stefan-Boltzmann Law), the temperature of the ground beneath an atmosphere that is opaque to infrared radiation is seen to be greater than the temperature of the ground when there is no such atmosphere present. This is the *greenhouse effect* of the atmosphere as it appears in this simple model of the earth-atmosphere system.

How much warmer is the ground temperature in this simple model compared to the case where no atmosphere is present? We already determined that the temperature of the ground in the latter case would be 276 K. Consider first the balance of radiation at the top of the atmosphere. In the steady state the incoming solar radiation of 342 W m^{-2} at (c) must be balanced by the outgoing infrared radiation emitted by the atmosphere at (f). This is the same balance as that for the model without an atmosphere. In other words, the temperature of the atmosphere must be 276 K.

Consider next the balance of radiation at the ground. The incoming solar radiation (c) plus the infrared radiation directed downward from the atmosphere (e) must equal the infrared radiation emitted by the ground (d). By application of the Stefan-Boltzmann Law we find that the temperature of the ground must be 328 K. In the model with no atmosphere, we determined that the ground temperature was 276 K. Thus, we have the result that the model with an atmosphere predicts an increase of 52 K (that is, 19 percent) in the ground temperature due to the greenhouse effect.

How does the result from our simple model compare to the observed greenhouse effect? As we earlier determined, the earth-atmosphere system radiates to space at an equilibrium temperature of 254 K, while the observed, globally averaged surface temperature is 288 K. Thus, the percentage increase due to the observed greenhouse effect is about 13 percent. Our simple model gives a result that is rather close to what is observed for the magnitude of the greenhouse effect.

A Nonisothermal Model

We now show that the isothermal model just explored does not embody any mechanism that can produce global warming. We begin by returning to Figure 2.6 in order to make a significant point. The arrow at (f) represents the radiation as if it were being emitted by the surface that constitutes the top of the atmosphere. Actually, the infrared radiation that the model atmosphere emits to space comes from inside the atmosphere. In other words, the effective radiating level (ERL) in this model is situated somewhere between the earth's surface and the top of the atmosphere. If we now add more greenhouse gas to the model atmosphere and render it more opaque to infrared radiation than it was, the altitude of the ERL will go up. If we add enough we can render the model atmosphere so opaque that the ERL will rise all the way to the top of the model atmosphere. (Conversely, a decrease in the greenhouse gas would lower the altitude of the ERL, and in the limit of complete absence of any greenhouse gas the ERL would descend to the earth's surface.) We can apply the same reasoning to the radiation emitted downward from the atmosphere at (e). There exists, in effect, an ERL for downward radiation, and the more opaque the model atmosphere, the closer to the surface this level will lie.

We explicitly recognize the existence of an ERL at some level inside the atmosphere by modifying Figure 2.6 to appear as in Figure 2.7. One other point is that the real atmosphere is *stratified*. In other words, the higher up you go in the atmosphere, the less dense (more rarified) the air. One consequence of the stratified nature of the atmosphere is that it is more opaque to infrared radiation near the earth's surface than it is higher up. We have incorporated this fact into Figure 2.7

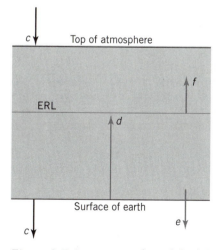

Figure 2.7 An improved model of the radiation balance of the earth-atmosphere system. Radiation emitted to space originates from the ERL shown in the diagram. For clarity, an ERL for radiation emitted downward from the atmosphere is omitted, although its location can be inferred from the origin of arrow (e).

by having the distance between the top of the atmosphere and the ERL for the radiation emitted to space at (f) to be larger than the distance between the surface and the level from which radiation is emitted downward at (e).

Now let's travel around in a few million cars burning up fossil fuels so that the amount of carbon dioxide in the model atmosphere goes up. Carbon dioxide is a greenhouse gas, so the model atmosphere will become more opaque to infrared radiation than it was, and the ERL depicted in Figure 2.7 will therefore go up in altitude. The outgoing radiation at (f) will not change, however, because our model atmosphere has the same temperature at all levels. There has been no change in the radiation balance of this model of the earth-atmosphere system; hence, no global warming ensues.

At this point, we will remove the assumption that the atmosphere is isothermal and assume instead that the temperature decreases with height at a rate equal to 6.5 K for every kilometer increase in altitude above the surface. This is the observed change of atmospheric temperature with height through the troposphere in the global average (Fig. 1.4).

We can use our earlier reasoning to understand what will happen if the amount of carbon dioxide increases in this more realistic model. As before, with the addition of new carbon dioxide, the ERL will rise to a higher altitude (Figure 2.8). Since the temperature of the atmosphere now decreases with height and the ERL is higher, the intensity of the infrared radiation emitted to space is less than it was before the new carbon dioxide was added. Now the radiation (f) emitted to space is less intense than the absorbed solar radiation (c), instead of being equal to it. In addition, the radiation emitted downward from the lower atmosphere is more intense than before. This is because the ERL for downgoing radiation has descended to a lower altitude, and there, in our atmosphere in which temperature decreases with altitude, it is warmer.

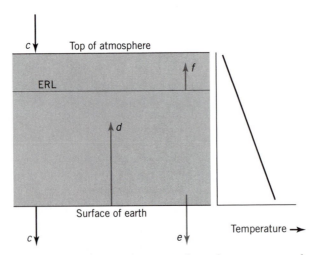

Figure 2.8 Same as Figure 2.7, but after more greenhouse gas has been added. The ERL for radiation emitted to space has risen. The variation of temperature with height in the model is shown at the right.

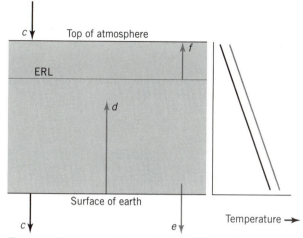

Figure 2.9 Same as Figure 2.8, but after the warming. The dashed line at the right indicates the new temperature profile.

Now we have an imbalance between absorbed and emitted (to space) radiation, and the earth-atmosphere system must respond by adjusting to a new steady state. The only way to reach a new equilibrium is for the atmosphere to warm until the temperature at the new altitude of the ERL (Fig. 2.8) is the same as it was at the old altitude of the ERL (Fig. 2.7). Recall now a basic property of this model of the atmosphere: The temperature is assumed to decrease with altitude at a rate of 6.5 K per kilometer. To maintain this property there must be an increase of temperature throughout the atmosphere exactly equal to the increase of temperature required at the new level of the ERL to restore radiative balance. Such a shift of the temperature profile is illustrated in Figure 2.9. After this adjustment, a new steady state is attained. In this new steady state the ground temperature is also higher than it was. It has to be, in order that radiation at (d) balance the sum of radiation at (c) and the increased radiation at (e).

2.3   THE EFFECT OF CLOUDS ON THE RADIATION BALANCE

Clouds and the Effective Radiating Level

Earlier in this chapter we calculated the ERL of the earth-atmosphere system in the global and annual average to be at an altitude of 5.2 km above the surface. This we called the *average* ERL. Its altitude is determined in part by greenhouse gases and in part by the presence of clouds in the atmosphere. In order to examine the effect that clouds have, we here define a *local* ERL that represents the ERL above some arbitrary place on the surface of the earth. Its altitude will, in general, not be the same as that of the average ERL.

The physical explanation for the greenhouse effect offered in the preceding section relied on a simple model of the atmosphere in which there were no clouds.

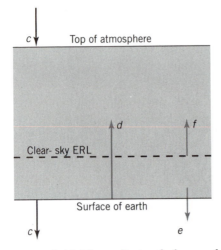

Figure 2.10 The radiation balance of the clear-sky model of the earth-atmosphere system with the clear-sky effective radiating level shown by a dashed line.

A greenhouse gas was assumed to be the only thing in the atmosphere absorbing infrared radiation. We now apply this model as a *clear-sky local model* of the atmosphere (Fig. 2.10).

When there are no clouds in the sky at our arbitrary place on earth, then the local ERL is what we will call the *clear-sky ERL*. Now we add some clouds to the picture and ask whether their presence makes the local ERL differ from the clear-sky ERL. Clouds are composed of droplets of liquid water and possibly crystals of ice and are totally opaque to infrared radiation. As a consequence of this total opacity, the infrared radiation emitted upward by a cloud comes from its top, not from somewhere down inside of it.

What then happens if some low clouds develop below the level of the clear-

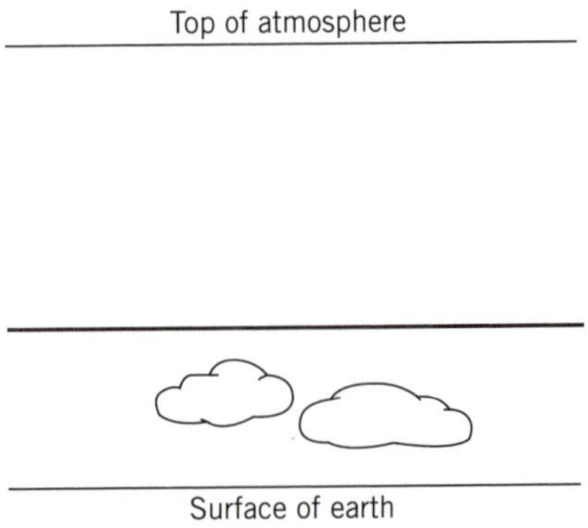

Top of atmosphere

Surface of earth

Figure 2.11 The ERL remains the same as the clear-sky ERL when low clouds are added.

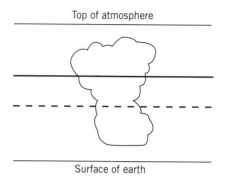

Top of atmosphere

Surface of earth

Figure 2.12 The ERL (solid line) is higher than the clear-sky ERL (dashed line) when a deep cloud is added.

sky ERL, as shown in Figure 2.11? In this case, there is no change in the height of the local ERL, since the clouds will emit radiation upward that will be absorbed by the atmosphere in the cloud-free regions above. This radiation will then not make it out to space, and the radiation that is emitted to space will be the same as it would be if the low clouds were not present; that is, the local ERL is unchanged.

What, then, happens if a deep cloud that extends through the depth of the troposphere develops, as shown in Figure 2.12? The local ERL must then be higher than the clear-sky ERL because the cloud radiates from its top, and the local ERL is an average of the clear-sky ERL and the cloud-top height. The average here is taken over an area that includes the cloud and the region of clear sky around it. Of course, if the cloud in Figure 2.12 covers the whole region we are considering, then the local ERL will be the cloud top.

Finally, we must consider what happens when we have a mixture of clouds with their tops at low, medium, and high levels as shown in Figure 2.13. Here, the answer depends on the relative mix of the three types and on the amount of clear sky in the averaging area, but we can at least see that the local ERL must lie above the clear-sky ERL.

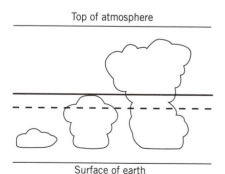

Top of atmosphere

Surface of earth

Figure 2.13 The ERL (solid line) is somewhat higher than the clear-sky ERL (dashed line) when a mixture of low, middle, and high clouds is added.

To summarize these possible effects on the local ERL of adding clouds:

- If the cloud top lies below the clear-sky ERL, then the local ERL will not change.
- If the cloud top lies above the clear-sky ERL, then the local ERL will move upward to a level that approaches the cloud-top height as the sky becomes more overcast.
- If the clouds are a mixture of high, middle, and low clouds, then the ERL will be higher than the clear-sky ERL.

We see then that, with the exception of clouds whose tops are below the clear-sky ERL, clouds act to *enhance* the greenhouse effect, since by raising the local ERL, they cause less radiation to be emitted to space. We expect that the earth-atmosphere system, given the existence of such clouds, is warmer than it would be without them.

The Albedo Effect of Clouds

We noted in Chapter 1 that clouds are the principal agent for reflection back to space of solar radiation. This property of clouds is referred to as the *cloud-albedo effect*. The cloud-albedo effect always acts to reduce the amount of solar radiation absorbed by the earth-atmosphere system and thus to make the system cooler than it would be if there were no clouds.

Clouds thus have two opposing effects on the radiation balance of the earth-atmosphere system. One is the greenhouse effect just described. If a cloud top is above the local clear-sky ERL, then that cloud is acting to reduce the infrared radiation emitted to space by the earth-atmosphere system and thereby make the system warmer. As we explained, the higher the cloud top, the greater the effect. But that same cloud will also reflect some visible solar radiation back to space and thereby act to make the system cooler than it would be if the cloud were not there. Unlike the greenhouse effect of the cloud, the albedo effect of the cloud does not depend on the height at which the cloud top is situated. Consequently, the lower the cloud-top height, the smaller the greenhouse effect relative to the cloud albedo effect. We will explore these two competing effects of clouds on the radiation balance in more detail in Chapter 7.

R E V I E W   Q U E S T I O N S

1. What (in the global and annual average) is the power received from the sun at the top of the atmosphere (in $W\ m^{-2}$), and what is the power absorbed by the earth-atmosphere system?
2. Why is the solar energy received and absorbed at the earth's surface less than the amount measured at the top of the atmosphere?
3. What is the equilibrium temperature of the earth-atmosphere system? What is the observed surface temperature of the earth in the global and annual average?

4. Why is the observed surface temperature of the earth 34°C higher than the equilibrium temperature of the earth-atmosphere climate system?

5. What are the conceptual differences between the average effective radiating level (ERL), the local ERL, and the clear-sky ERL?

6. If the atmosphere is assumed to be isothermal as in the simple model discussed in section 2.2 and the amount of carbon dioxide in the atmosphere increases, why does the temperature remain unchanged?

7. With the aid of Figure 2.7, describe the exchange of radiation between the earth, the atmosphere, and space.

8. What is the greenhouse effect, and how is it produced?

9. If temperature decreases with altitude, an increase in the altitude of the clear-sky ERL produced by an increase in greenhouse gases results in a decrease in the infrared radiation emitted to space. How does this result in global warming?

10. How does the albedo effect of clouds differ from the greenhouse effect of clouds?

11. Imagine for the moment that global warming results in temperature increasing so much that low clouds are formed less frequently than at present. Could this reduction of low clouds contribute to further global warming?

INTERNET COMPANION

2.1   Calculating the Radiation Balance

- The CoVis greenhouse visualizer developed at Northwestern University provides considerable information on the radiation balance of the earth-atmosphere system. Information on the albedo and longwave radiation emitted to space are provided.

2.3   The Effect of Clouds

- Time-lapse photos and short video clips of clouds from a camera located on the University of Utah are available. Views of clouds from many other locations around the country are also available on the University of Michigan WeatherNet Weather Cam page.

- A fact sheet on the role of clouds in the radiation balance is available from NASA. This information includes color schematics similar to those presented in this section.

CHAPTER 3

Weather and Climate

In Chapter 1 we described how observations of surface temperature have been processed to constitute a record of climate over the past 130 years. We made explicit the idea that when we are talking about climate, we are talking about quantities that have been averaged over a period of time. With this in mind, we now define weather as "what is going on in the troposphere at any particular time at any given place on the globe." Climate is then the weather averaged in time over some specified period. Weather is described by quantities such as temperature that we can measure and try to predict. Climate is described by the time average of such quantities.

A fundamental characteristic of weather is air motion, which we measure and forecast as wind. We begin by addressing the question of why the winds blow and by looking at the operation of the atmosphere as an engine. Then we consider satellite cloud photography as an example of a global weather observing system. How weather is predicted by means of numerical models is then treated, and the manner in which such models are used for climate prediction is considered. The chapter concludes with an analysis of the hydrologic cycle.

3.1 THE WEATHER

Why Do the Winds Blow?

The answer to this question, from the perspective of this book, is that the winds (and also the currents in the ocean) maintain the radiation balance of the earth-atmosphere system. In the previous chapter, we dealt with the global-average system, in which there are only two directions: up to space and down to the earth's surface. While this is sufficient for understanding the greenhouse effect, neither the intensity of solar radiation absorbed nor the intensity of infrared radiation emitted to space is the same at all points on the globe. In particular, there is a strong dependence on latitude.

The sphericity of the earth affects the amount of sunlight received over a given patch of the earth in a way that is illustrated in Figure 3.1. Because of the great distance between the earth and the sun, the solar radiation that is intercepted by the earth's surface is traveling along the parallel rays depicted in the figure. The power carried between any two rays spaced an equal distance apart is equal. The figure as drawn thus shows three equal "bundles" of solar radiation incident in each hemisphere. It is evident in the figure that the bundle incident in high latitudes is received by a larger surface area than the bundle received in tropical latitudes. This is a consequence of the fact that in high latitudes the receiving surface is more inclined to the ray. Recall that intensity is defined as power received divided by the area of the surface that receives it. It follows that the intensity of solar radiation received at the surface decreases with latitude.

The heavy line in Figure 3.2 shows the variation with latitude of the annual-average intensity of solar radiation absorbed by the earth-atmosphere system. This is visibly greater than the global average (237 W m^{-2}) in low latitudes and less than the global average in high latitudes. Also shown in Figure 3.2 is the variation of the annual-average intensity of the infrared radiation emitted to space. The intensity decreases with latitude less rapidly than does the intensity of absorbed solar radiation. This decrease is a consequence of a decrease with latitude of the temperature at the level at which the ERL is situated.

The curves in Figure 3.2 divide the earth-atmosphere system into (1) the regions poleward of about 35° in each hemisphere where absorbed solar radiation is less than infrared radiation going out to space, and (2) the region equatorward of about 35° where the opposite situation prevails. There is thus a radiation imbalance such that energy is supplied to (lost from) the earth-atmosphere system equatorward of (poleward of) about 35° latitude. Yet, in the annual average, the system is not observed to heat up and cool down in the respective regions. What is occurring is that, in the course of a year, the atmosphere and the ocean transfer poleward the amount of energy that is needed to offset the radiation imbalance

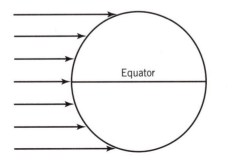

Figure 3.1 Solar radiation incident on a spherical earth.

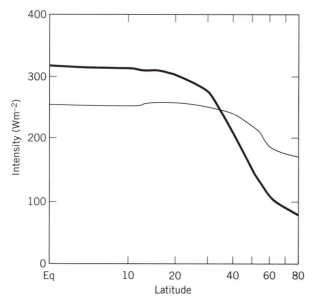

Figure 3.2 The variation with latitude of absorbed solar radiation (heavy line) and of infrared radiation emitted to space (light line) in the annual average. Units are W m^{-2}. The scale of latitude is such that equal lengths along the scale correspond to equal areas on a spherical earth.

completely. We thus have an answer to the question posed: The winds must blow in order to transfer energy poleward. We next consider what sort of wind pattern would do this.

The Atmospheric Heat Engine

The radiation imbalance just described constitutes for the earth-atmosphere system a source of heat in low latitudes and a "sink" of heat in higher latitudes. What the atmosphere does in this environment is to function as a heat engine. Considered this way, the atmosphere does three things: (1) it absorbs heat, (2) it converts part of this heat into motion, and (3) it gives up the remaining part of the heat as exhaust. A more familiar example is a car engine. It acquires heat from fuel combustion, it converts part of this into motion, and the exhaust gas carries off the remaining part of the heat. The term "efficiency" applied to a heat engine refers to the part of the heat acquired that gets converted into motion. The atmosphere is a very inefficient heat engine; only a very small part of the heat that it takes in is converted into motion.

The heat-intake and heat-exhaust aspects of the atmospheric engine are determined by the radiation balance of the atmosphere considered as a system in its own right. What we showed in Figure 3.2 was the variation with latitude of the radiation balance of the earth-atmosphere system (the atmosphere and the ocean). The variation with latitude of the radiation balance of the atmosphere alone, which we do not show here, is quite similar. All we need to know for our purposes here is that tropical latitudes are the scene of the heat intake of the atmospheric engine and high latitudes are the scene of the heat exhaust.

Let us put forward now a relatively simple pattern of global motion that would prevent this distribution of heating and cooling from resulting in an ever-increasing temperature in tropical latitudes and an ever-decreasing temperature in high latitudes. The argument is based on the principle that air being heated will rise and air being cooled will sink. We represent this distribution of rising and sinking with the vertically oriented arrows in Figure 3.3. Such motion can of course be maintained only if the air that is rising is replaced and the air that is sinking toward the surface is evacuated. Both requirements are satisfied by equatorward flow in the lower atmosphere, as represented by a horizontal arrow in Figure 3.3. The same reasoning leads to the poleward flow in the upper atmosphere represented by the other horizontal arrow in this figure. The equatorward flow brings air from the region of cooling to the region of heating, and the poleward flow brings air from the region of heating to the region of cooling. Both flows are achieving a poleward transfer of heat.

The pattern in Figure 3.3 is a circulation cell extending from the equator to high latitudes with a circulation directed clockwise. This figure is for the Northern Hemisphere. If we were to do this figure for the Southern Hemisphere, with equatorial latitudes (and the associated ascending arrow) located on the right hand side of the figure, then the resulting circulation would be directed counter-clockwise.

Such a pair of equator-to-pole cells does exist in the atmosphere, but the flow in them is so weak that it is not observable. More specifically, it is completely swamped in the troposphere by the much more vigorous and varied circulations producing daily weather. The best means of viewing these circulation patterns in a global framework is a mosaic of satellite photographs of clouds. We show one of these in Figure 3.4. Here, clouds act as tracers and reveal the global field of motion in the troposphere on a particular day. The dominant features everywhere except near the equator are large rotating swirls. A sequence of photographs like this would reveal that these swirling patterns move from west to east. These patterns are the middle-latitude storm systems. Any particular one loses its identity

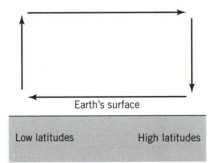

Figure 3.3 Schematic of an atmospheric circulation that will provide a poleward transport of heat.

Figure 3.4 Clouds as seen from space on May 15, 1995. Notice the large swirling cloud mass over the North Atlantic Ocean.

in less than a week and is replaced by a new one somewhere else. This endlessly changing pattern of moving storm systems is what accomplishes almost all the poleward heat transfer that takes place in the atmosphere.

What we have disclosed here is that almost all the motion produced by the atmospheric heat engine is of a far more complicated character than that of the circulation cell depicted in Figure 3.3. What is ultimately responsible for the atmosphere choosing the more complicated option is the rotation of the earth about its axis. How and why this results in motion like that revealed in Figure 3.4 is the subject of that part of the field of meteorology known as dynamics and would require more explanation than is necessary to pursue here.

The active weather systems apparent in Figure 3.4 are confined to the troposphere. The stratosphere is a much more tranquil region—so much so, in fact, that the upper part of the circulation cell in Figure 3.3 comes into prominence there as a means to transport stratospheric ozone from the equatorial latitudes, where it is produced, to the polar regions (of both hemispheres), where it is stored. We shall return to consider this aspect of global atmospheric motion in Chapter 9.

3.2 SATELLITE CLOUD IMAGERY

Time-lapse sequences of cloud photographs taken by cameras mounted on geostationary satellites have become a mainstay of weather visualization. Everyone has seen this imagery in the weather segments of television newscasts. What you may not have realized is that what is usually shown is an infrared image. That is, the radiation that the camera is seeing is infrared radiation emitted by the earth-atmosphere system. There are situations in which visible rather than infrared imagery is also useful for visualization. In this section we look at what these two types of images tell us.

The satellites fall into two categories: geostationary satellites and polar orbiting satellites. Geostationary satellites remain fixed in position over a point on the equator. By remaining fixed with respect to the earth's surface, they can monitor the pattern of clouds continuously in time. Polar orbiting satellites do not remain fixed in position over a point on the earth's surface. Instead, they fly along an elliptical orbit which passes almost over the poles. Each such satellite achieves near-global coverage because the earth is rotating inside the satellite's orbit. This makes the polar orbiting satellite ideal for monitoring the radiation balance of the earth-atmosphere system. We will return to polar orbiting satellites in Chapter 7, when we examine the separate contribution that clouds and greenhouse gases make to the radiation balance.

The instruments on board a geostationary satellite are cameras similar in a gross sense to surveillance cameras operating in security systems for homes and businesses. One type of camera operates in visible light. We will refer to the output from this sort of camera as a *visible image*.

Like the observer on the ground attempting to look up at the clouds at night, a satellite with a camera pointing down that is sensitive only to visible light is not able to see very much. Figure 3.5, left-hand photo, shows a visible image taken at 8 P.M. Mountain Daylight Time (MDT) on August 20, 1993. Notice that a hurricane (Hurricane Greg) can be seen dimly near the bottom left of the figure. The hurricane is distinctive in its nearly circular hole (the eye). To the east of the hurricane, the image fades out. Are there any other storms in those regions? We can't tell because the sun has set below the horizon there.

Surveillance cameras at commercial establishments are not restricted to the visible part of the electromagnetic spectrum. Those concerned with nighttime security outside the building use cameras that see in the infrared. Weather satellites are another use for this technology. Figure 3.5, right-hand photo, shows an infrared image taken at the same time as the visible image shown in Figure 3.5, left-hand photo. The region east of the sunset line is no longer in the dark for us.

Before we can interpret all that such an infrared image can tell us, we must go a little deeper into the details of the infrared radiation that the earth-atmosphere system emits. Up to this point, we have been able to get away with defining a greenhouse gas as one that absorbs *all* infrared radiation emitted by the surface of the earth. In reality, the amount of absorption depends on the wavelength of the radiation. Thus, it turns out that there are some parts of the spectrum of infrared radiation where absorption by greenhouse gases of the infrared radiation emitted by the surface of the earth is incomplete. Those parts of the infrared

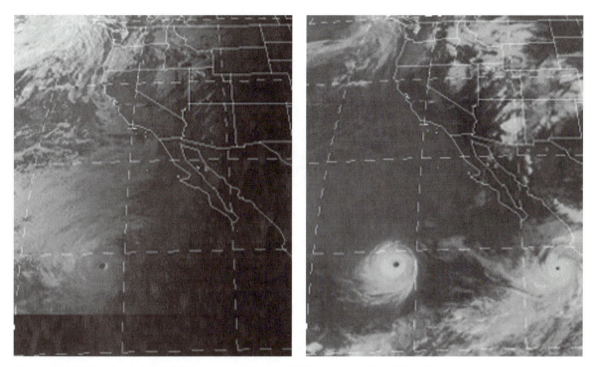

Figure 3.5 Left photo: A visible image at 8 P.M. MDT, August 20, 1993. The sun has set for locations in the eastern half of the image. Hurricane Greg is located in the lower left quadrant of the image. Right photo: Infrared image at the same time. Hurricane Hilary is located in the lower right of the image.

spectrum are termed atmospheric windows, and we refer to the infrared radiation in these parts of the spectrum as *window radiation*. These parts of the spectrum are illustrated in Figure 3.6. To use this figure, we first select a wavelength and then trace a line vertically upward to the edge of the black. If that edge is at 100 percent, then the radiation at that wavelength is completely absorbed (that is, is not transmitted through the atmosphere) by the gases that are present in it. Those wavelengths in the infrared spectrum for which the edge of the black is situated down around 10 or 20 percent constitute atmospheric windows. For example, infrared radiation with a wavelength of 10 μm is considered window radiation.

We are here considering only the absorption of infrared radiation by atmospheric gases. Clouds are not atmospheric gases. They are composed of small liquid drops or small crystals of ice. We can stick with our assertion in Chapter 2 that clouds are completely opaque to infrared radiation. They thus absorb all infrared radiation emitted from the earth's surface, and they radiate infrared radiation upward from their tops. It is quite adequate for our purposes in this book to assert that this statement is valid for all wavelengths of infrared radiation.

It turns out that window radiation makes up a rather small fraction of the total infrared radiation emitted to space by the earth-atmosphere system. What this means is that we can ignore its existence when looking at the basic processes

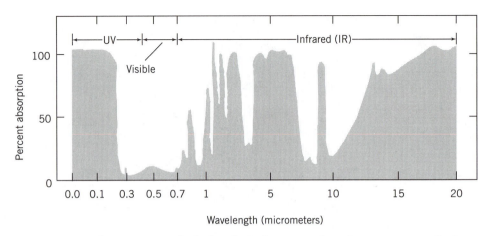

Figure 3.6 The extent to which the clear-sky atmosphere is opaque to radiation at different wavelengths.

that determine its radiation balance. That is what we have done in Chapter 2, where the simple models advanced were founded on the definition of a greenhouse gas as one that absorbs all the infrared radiation emitted by the earth's surface.

Infrared-image cameras in satellites are tuned (filtered) to receive only window radiation. But window radiation, as the preceding discussion hopefully makes plausible, is infrared radiation whose local clear-sky ERL is essentially down at the surface of the earth. By local clear-sky ERL, we mean the local ERL in a cloud-free region. As a result, what is seen in the infrared images from satellites are cloud tops and, between the clouds, the surface.

There is a further advantage of photography using window radiation rather than total infrared radiation, and that is one of clearer background contrast. This is a point that we can better make after looking at some examples. Figure 3.7 shows visible and infrared images taken at noon MDT on August 20, 1993, eight hours before the images shown in Figure 3.5. Let's look at the visible image first (the upper lefthand frame). The black areas to the west of Baja California are places where there is little sunlight reflected back to space because the skies are clear and the ocean absorbs most of the visible light. Highly reflective clouds spread northward along the Arizona–New Mexico and Utah–Colorado borders. Notice that Hurricane Greg is much more distinctive in the noon-time image than it is in the one taken in the evening (Fig. 3.5a). Also apparent at mid-day is another swirl of highly reflective clouds near the west coast of Mexico. This is another developing hurricane, Hilary.

An infrared image taken at the same time is shown in the upper righthand corner of the figure. The convention here is the same as that in the visible image: Areas of low-intensity radiation appear black and areas of high-intensity radiation appear white. Let us recall now what governs intensity of radiation. The intensity of the reflected radiation that is seen in the visible image depends on the character of the earth's surface at any given point and on whether a cloud is present there. The intensity of emitted infrared radiation, on the other hand, depends on the

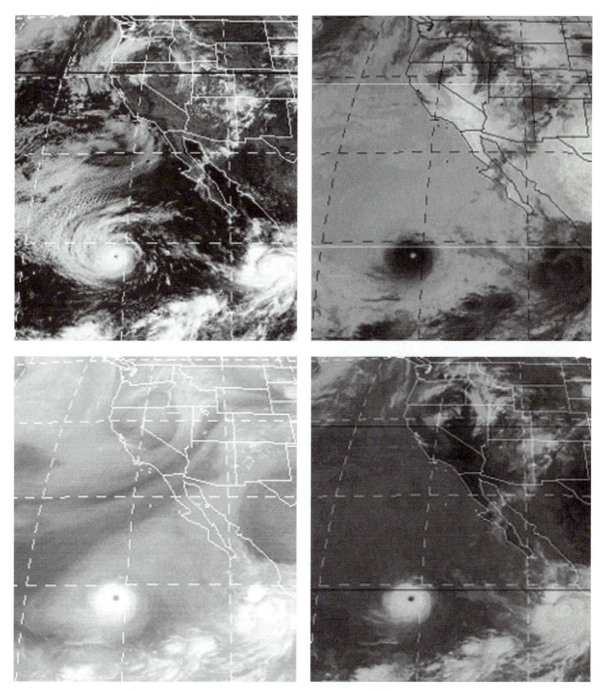

Figure 3.7 Images taken at noon MDT on August 20, 1993, in different parts of the electromagnetic spectrum. Upper left: visible image. Upper right: infrared window-radiation image. Lower right: the negative of the same image. Lower left: the negative of an infrared radiation image with wavelengths near 6.7 μm.

temperature of the emitting object in accordance with the Stefan-Boltzmann Law that was introduced in Chapter 2. Thus, the surface of the earth appears white, and the clouds defining Hurricane Greg, whose tops are quite high up in the atmosphere where temperature is much lower than it is at the surface, appear as a black area.

Comparing the infrared and visible images, it is possible to see areas that are covered by highly reflective clouds that do not have cold cloud-top temperatures. These are low, relatively warm clouds found over the ocean to the west of California. They are emitting nearly the same intensity of infrared radiation to space as is the underlying surface. It is much more difficult to distinguish between low clouds and clear regions in the infrared image than it is in the visible image.

We are not used to thinking of clouds as black. There is an easy trick for rendering infrared photographs more intuitively appealing in this respect. We simply use the negative of the infrared photograph instead of the positive that is seen in the upper righthand panel of our figure. We shall henceforth refer to such a negative as the *infrared image*. The lower righthand panel shows the infrared image in this example. Hurricane Greg and other masses of clouds with high tops appear white. Clouds with lower tops appear grey. In this infrared image we can even detect that the land surface in southern California and adjacent regions (black) is warmer than the surface temperature of the Pacific (dark gray).

The bottom lefthand panel of the figure is an infrared image taken with a camera filtered to admit only infrared radiation with wavelengths in a restricted part of the spectrum near 6.7 μm. This, as can be seen from Figure 3.6, is *not* window radiation. Where there are clouds with high tops, as in Hurricane Greg, the radiation that we see is being emitted from the cloud tops. Elsewhere it is being emitted from the clear-sky ERL of the greenhouse gas water vapor. This narrow part of the infrared spectrum that the camera filter is admitting is for this reason known as the "water-vapor channel." Such an infrared image enables us to see how water vapor is spatially distributed, which is important since the number of water vapor molecules is several orders of magnitude larger than the number of any of the other greenhouse gases, including CO_2.

When we compare the two lower panels in our figure, we see that the water-vapor channel offers a rather poor background contrast for infrared imagery of clouds. Infrared imagery taken using the total (all-wavelength) field of infrared radiation emitted by the earth-atmosphere system looks very much like that taken in the water-vapor channel. The resulting background contrast is poor, and what's more, low cloud tops don't show up at all. This is the other advantage we alluded to earlier of using window radiation in satellite photography. The contrast is good, and we don't miss out on the low clouds.

3.3 THE PREDICTION OF WEATHER AND CLIMATE

An interesting paradox exists in the public's perception of the prediction of weather and climate. As individuals, we may have a healthy distrust of the accuracy of the forecasts by local National Weather Service forecasters and television broadcasters of what the weather will be like tomorrow. These forecasts are based to a large degree on simulations of the atmosphere using models. However, we generally accept the claims made about global warming that are based on similar models that are used to simulate the atmosphere for decades into the future. So the paradox is simply this: How can you trust a model of the atmosphere to predict the climate as much as a hundred years in advance if you don't trust similar models to predict the weather tomorrow?

To some degree, we *can* have confidence in a simulation of the climate a hundred years from now while still having doubts regarding the accuracy of the forecast for tomorrow simply because the prediction of climate is quite a different matter from the prediction of weather. A weather prediction needs to be very accurate in order to know whether or not it will snow tomorrow in Salt Lake City, for example. On the other hand, the prediction of climate requires less accuracy and detail. We are interested then in whether the model can generate a realistic average of all the individual weather systems. That is to say, it isn't important exactly where or on what day the snow falls, as long as it snows a reasonable amount over the country as a whole during winter.

The Workings of a Weather Prediction Model

Let's begin our discussion from the perspective of what is required to make a prediction of the weather at Salt Lake City for tomorrow. The steps required to issue such a forecast are as follows:

- Observations of the temperature, winds, clouds, pressure, and so on, are made around the globe by participating governments.
- These observations are processed so that the global state of the atmosphere at a particular time is specified as well as possible.
- Models of the global atmosphere developed by the National Centers for Environmental Prediction of the National Weather Service are used to predict the state of the atmosphere up to 10 days in advance.
- These forecasts of the atmosphere are interpreted by personnel of the National Weather Service, and their expectation of what the weather is likely to be at Salt Lake City is disseminated over radio and television.

Now, let's discuss each of these steps in more detail. All the governments around the globe cooperate by exchanging weather information through an entity called the World Meteorological Organization. This normal exchange of weather information proceeds quite successfully except when regional strife develops. Countries at war do not exchange weather information nor do civil wars provide an environment in which people have time (or inclination) to measure the temperature.

The information required to start a model of the atmosphere includes not only measurements of temperature, wind, and so on near the earth's surface around the globe but also detailed information on the vertical distribution of these quantities. For this purpose, balloons are launched twice a day around the globe to determine the state of the atmosphere in three dimensions. Satellite-derived measurements of cloudiness, temperature, and winds are also used.

The process of taking a mix of different types of observations at scattered points around the globe and determining the current state of the atmosphere brings up problems similar to those alluded to in Chapter 1, where the objective was to determine the globally averaged air temperature from observations scattered around the globe on land and at sea. However, determining the three-dimensional global structure of temperature, wind, pressure, and moisture at an instant is a problem vastly more complicated than that of deducing the globally averaged surface temperature.

The basic problem is to determine from irregularly spaced observations the appropriate values of temperature and other weather variables at a lattice of points in three dimensions as shown in Figure 3.8. The points that we are referring to here are defined by the intersections of the lines that constitute the lattice.

Notice that the lattice follows the terrain of the underlying surface in order to be able to simulate the atmosphere over regions of high terrain such as the western United States. A major consideration in making a weather prediction is to determine the horizontal and vertical resolution that the lattice should have. The lattice in Figure 3.8 has only three levels in the vertical, and these are spaced about equally. The actual lattices for such forecasts are constructed for as many as 50 levels in the vertical and hundreds of points in each direction in the horizontal extending over the whole globe.

The global models of the atmosphere that are used to predict weather are much more complex than the simple models of the atmosphere developed in the previous chapter. These *numerical weather prediction (NWP)* models are based

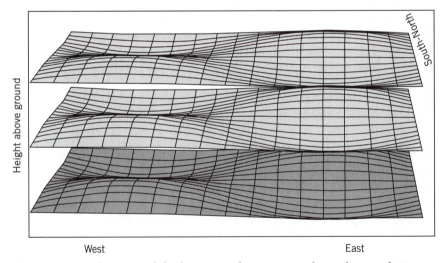

Figure 3.8 A schematic of the lattice used in numerical weather prediction models. The lower surface represents the terrain of the earth while the two other surfaces represent other levels on which the state of the atmosphere is predicted. The temperature, wind, and pressure would be determined for each grid intersection on each level.

on a set of equations that describe the fluid motion of the atmosphere. These equations relate how the temperature, wind, pressure, and moisture fields are determined from the physical forces that cause motion in a fluid, such as the pressure gradient force, gravity, and the force that arises from the rotation of the earth about its vertical axis.

While the equations that govern fluid motion in the atmosphere are accurately known, their actual solution by computer is quite involved. This is because we have equations valid at each point of the three-dimensional lattice and each equation takes into account the interactions that take place between this point and adjacent points of the lattice. The total number of equations that must be solved by a model of the global atmosphere can easily exceed several tens of thousands.

To understand why all of these equations are interrelated, consider the schematic in Figure 3.9 of a few adjacent points of the lattice over which a westerly wind is blowing. Given that the points upstream are colder than the points downstream, the westerly wind will carry the cold air to the downstream points. The equations for temperature at the downstream points of the lattice must therefore include the temperature of upstream points as well as the wind that blows across them. In this manner, the characteristics of the atmosphere in terms of temperature, moisture, wind, and pressure, are *advected* (that is, moved) around the globe.

Besides simply knowing the equations that govern the motion of fluids, we must be aware of other characteristics of the model that affect how skillful our forecasts will be. These include the horizontal and vertical resolution of the model's lattice and the length of the time interval between computations of the variables of the model. As a general rule, as we increase the spatial and temporal resolution of the atmospheric model, we will improve the quality of the model's forecast of the weather. However, we cannot use as high a spatial resolution as we might like because the number of equations then becomes so large that they

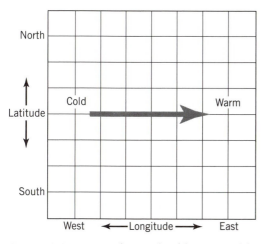

Figure 3.9 A westerly wind is blowing cold air toward warm air across the lattice of a numerical weather prediction model.

cannot be solved in a reasonable time. In addition, if we reduce the size of the time step between computations of the model variables, then the model has to compute the equations more often.

Because of these factors, the size of the computer on which the model is run becomes a limiting issue. Atmospheric scientists have always relied on the biggest computers available in terms of memory and speed. Since there are several tens of thousands of equations that comprise a model of the global atmosphere, the computer must be able to process all of these equations in an efficient manner, or else so much time will have elapsed that the model will end up forecasting the atmospheric state for a time that has already past. Thus, as computers become faster, the models of the atmosphere can become more complex in terms of the number of equations that can be solved. They then become more accurate in predicting the weather.

After a model completes a forecast, it then is possible to view how the model forecast has evolved in three dimensions and in time. Time-lapse sequences of graphic model output in which the evolution of the entire model forecast is stepped through rapidly then given a visual impression of how the atmosphere is expected to evolve in the model. Atmospheric scientists in the federal government and in private industry use such model output to interpret how weather conditions are likely to evolve at a specific location.

Although output from such prediction models can be used directly to tell someone what the surface temperature would be at a given place, this direct model output is always in error. For example, topography-resolution problems characteristically place Salt Lake City several hundred meters higher in altitude than it really is. Because of such *model biases*, the model output must be interpreted by local forecasters in order to improve the quality of the forecasts issued to the public. In addition, the local forecaster incorporates information that is not available to the model, such as the forecaster's own experience with the local weather.

Why Are Weather Predictions Often Wrong?

It should be clear from this discussion that there are plenty of reasons why the forecast heard on the radio may be entirely wrong. How many times have you heard as you are traveling in a downpour on a drenched freeway: "Today will be party cloudy"? Part of this can result from the human component. The radio station may be broadcasting an old forecast issued 6 to 12 hours earlier, and conditions have changed rapidly. However, in most cases, the errors are a result of the imperfect nature of the models' simulations, independent of any direct errors by people.

As mentioned before, the models of the atmosphere represent the state of the science in terms of incorporating the physical factors that we believe to be important for the prediction of weather. Thus, some of the errors result from those occasions when factors that are not adequately handled by the models arise or assumptions that are made in the model are invalid.

If we could construct a perfect model of the atmosphere that incorporated all the physical forces that cause motion in the atmosphere, would we be able to forecast the weather perfectly? Surprisingly, the answer is no.

There are some physical systems for which it is possible to predict in advance what is going to happen. You can obtain tide tables for San Francisco Bay for the next hundred years, and the paths of the planets around the sun can be predicted far in advance. Unfortunately, the atmosphere is not such a system. There is a limit beyond which it is not possible to predict the weather accurately. The generally accepted estimate for this limit is somewhere between 12 and 15 days. This limit arises from the *chaotic* nature of the atmosphere. Even if we had a perfect knowledge of the workings of the atmosphere, we can never observe and measure the atmosphere to the degree of accuracy necessary to predict the details of its future beyond this limit. Edward Lorenz originated an explanation for this phenomenon that has penetrated down to the elementary school level of education, which can be paraphrased loosely as follows: The undetectable flap of a butterfly's wings in the Amazon Basin introduces enough uncertainty into a model of the atmosphere that it eventually destroys the accuracy of its forecast over the United States.

The schematic in Figure 3.10 shows a fictitious record of air temperature at Salt Lake City over a 12-day period. The temperatures rise and fall as a result of weather systems that cross the area. Now imagine that we have a model of the

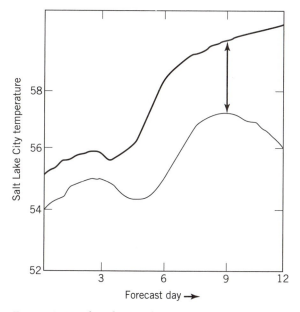

Figure 3.10 The observed temperature at Salt Lake City (thin line) and the predicted temperature (heavy line) as a function of time. The initial error of the predicted temperature was 1°F. The vertical line with the double arrows indicates the point at which the error of the forecast becomes larger than that of a climate forecast using the climate normal for that day.

atmosphere for which the initial state of the atmosphere on day 0 is known quite well, except for the temperature at Salt Lake City (for argument's sake assume that the observer mistakenly sent in the wrong temperature by 1°F). Then over the next 12 days, the temperatures forecast at Salt Lake City by the model might evolve as shown. At the initial time, there is the 1°F error, and this grows later to more than 4°F. We can also imagine that this unintended error introduced by the observer at Salt Lake City slowly affects the forecast over the entire United States. Eventually the error at Salt Lake City becomes so large that the forecast is useless (indicated in the schematic as day 9). At this point, the difference between the observed temperature and the forecast temperature is so large that it is greater than the difference between the observed temperature and the temperature normal (see section 1.2) for that day. In other words, we would be better off using the climatic normal to estimate the temperature beyond day 9 in our example rather than the output from our model of the atmosphere.

The growth of the error at Salt Lake City is a crude example of the chaos phenomenon that originates from the advection process described earlier. Advection is a *nonlinear* process that allows the weather at one point to cause chain reactions that eventually affect the weather at many other points.

The chain reaction aspect of advection over the globe would not contribute to chaos if it weren't for the fact that it is a nonlinear sequence of events. Consider Figure 3.11 in which a *linear* equation is plotted along with two nonlinear equations. The equations themselves have no importance, but what we would like to know is, if x changes by a fixed amount, what happens to y? Focus for the moment on the lower figure. For the linear equation (thin solid line), y and x change by equal amounts. If we move from 1 to 1.2 in the x direction, then we move from 1 to 1.2 in the y direction as well. Now, what happens for the other two equations? As x changes from 1 to 1.2, y changes from 1 to 1.4 (thin dashed line) or from 1 to 2 (heavy line). In other words, a nonlinear process can cause small changes to grow into large ones. The heavy dashed curve is similar to the Stefan-Boltzmann Law (see section 2.1). Remember that we showed that if the temperature doubled, the energy emitted by a body would increase by a factor of 16. We can see the same thing in the upper figure. Note that as x varies from 1 to 2, y changes from 1 to 16 in the heavy curve.

A linear system can be defined for our purposes as one in which the changes introduced at an early stage of the system are passed through to its later stages without amplification. The linear version of Lorenz's butterfly example could perhaps be the following: The undetectable flap of a butterfly's wings in the Amazon Basin eventually causes an undetectable change in the wind over the University of Utah. Or for the case of our example of the temperature at Salt Lake City, the error would remain constant with time, equal to 1°F. So, in a linear system, small errors at the initial stages of a model's forecast will not grow.

A nonlinear system differs in that the change at early stages in the system may magnify (or diminish) with time. The nonlinear process of advection in the atmosphere eventually over time becomes so overwhelming and ultimately affects the accuracy of the forecast over the entire globe because there are thousands of grid points involved in millions of nonlinear computations.

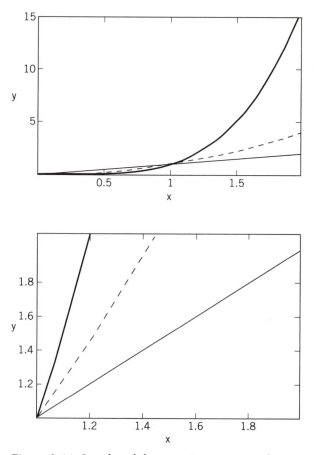

Figure 3.11 Graphs of the equations $y = x$, a linear equation (thin solid line), $y = x^2$, a nonlinear equation (thin dashed line), and $y = x^4$, another nonlinear equation (heavy line). The lower graph shows in greater detail a section of the upper one.

The Workings of a Climate Model

The best model used to simulate the climate of the atmosphere for extended periods does not differ substantially from a model used to predict the weather for a few days. Both types of simulations suffer from the chaos phenomenon that arises from nonlinear advective processes. A climate model has no value to predict the weather 10 years in advance; the best prediction of the weather on a specific day that far in advance is to use the climate normal (as defined in section 1.2) for that day.

The value of the climate model arises from its use to evaluate specific physical processes in the atmosphere and the sensitivity of the climate to these processes. The simple models developed in section 2.1 were adequate to conceptualize the global warming, but for the best estimates of climate change, we must use a climate model that computes the interactions between the atmospheric variables around the globe.

There are a number of procedural differences between weather and climate models. We stated earlier that a weather prediction model is improved if the horizontal resolution is increased. On the fastest computers available at the present time, a 10-day prediction of the weather takes roughly 4 hours to complete. If we were to run this same model for 100 years, it would take slightly less than 2 years to finish. Because computers normally do several tasks at once, it would actually take several times that length of time to finish. Thus, we can't afford either the time or the expense to use the high horizontal resolution of a weather prediction model for climate applications. Figure 3.12 shows that the typical climate model has relatively poor horizontal resolution as evidenced by the coarse boundaries between the oceans and land areas.

The numerics of atmospheric models allow the time interval between predictions of the variables of the model to be lengthened when the horizontal resolution is degraded. The net effect of the reduced horizontal resolution and longer time step is that a climate model takes only a few minutes to complete a 10-day forecast. Assuming that a climate model takes 1 minute of computer time for each model day, a hundred-year run could be completed in roughly 2 months. This is a reasonable length of time for computations as part of a multiyear research project.

Although the resolution of a climate model is by necessity degraded in order to complete its prediction in a reasonable amount of time, many other aspects of the model must be vastly improved compared to a weather prediction model. We have already seen that the earth-atmosphere climate system requires an understanding of not only the atmosphere but also its interactions with the underlying surface. In a weather prediction model, the surface is usually treated as remaining fixed over the length of time of the forecast; in other words, the sea-surface tem-

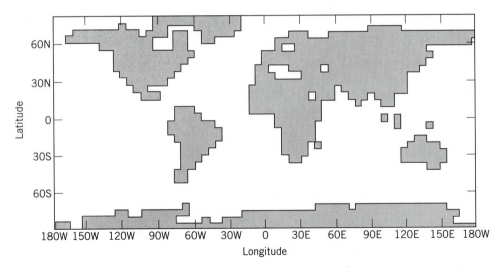

Figure 3.12 The degraded horizontal resolution of climate models implies that their depictions of land and sea boundaries are usually quite crude.

perature of the ocean and the amount of ice over the polar regions remain fixed. However, this is not possible in a climate model that hopes to capture the evolution of the climate system adequately over extended periods. For this reason, climate models must predict the evolution of not only the atmosphere but also the ocean. The requirement that climate models simulate the coupled ocean-atmosphere system introduces considerable complexity into the simulations. It also requires vastly expanded computer resources.

3.4 THE HYDROLOGIC CYCLE

While we have in this text adopted the global-average surface temperature as the index of global warming, the local changes in precipitation that may accompany global warming are likely to have the greatest impact on humanity and on the global ecosystem in general. What we will do here is to look briefly at the cycling of water through the global atmosphere. Water evaporates globally from the surface, enters the atmosphere as water vapor, and then precipitates out. This global cycle is known as the *hydrologic cycle*. Its treatment at this point in the book enables us to introduce the idea of *residence time* that we will use later to look at other global processes.

Residence Time

Consider a bathtub filled with 50 gallons of water. We open the drain and water flows out at a rate of, say, 10 gallons per minute. It is easy to see that the time it takes for all of the water to run out is 5 minutes. We say that the *residence time* of the water in this *reservoir* (the bathtub) is 5 minutes.

Now we consider the same numerical example, but this time as we open the drain, we also turn on the tap, so that water runs in as well as out at 10 gallons per minute. The result is that 50 gallons of water will remain in the tub. We say in this situation that the reservoir is in a steady state. The residence time of the water, which we defined as the amount of water in the reservoir divided by outflow rate, will be 5 minutes, as before. For this or any example of a steady state, because the flow into the reservoir equals the flow out, we could just as well define residence time as reservoir content divided by the inflow rate.

Residence Time of Water in the Atmosphere

We now consider the global atmosphere as a reservoir of water. Most of the water present there is in the form of water vapor, which is a gas. Water in the form of liquid droplets and ice crystals, which we perceive as clouds, constitutes a small percentage of the atmosphere's water content. Balloons are released every day at weather-observing stations over the globe. These carry sensors that measure temperature and the amount of water vapor (humidity) as they ascend through the troposphere. The global and annual averages of these temperature data at each level have been used in determining the temperature profile that was shown in

Figure 1.4. In like manner, the humidity data can be used to define the global- and annual-average water vapor content at all levels in the atmosphere above the surface. From this information the global- and annual-average amount of water (in the form of water vapor) in the atmosphere can be calculated. How much water does this correspond to? If we were to condense it all instantly as a liquid, it would amount to a layer about 1 inch (2.54 cm) deep distributed over the whole of the globe. The area of the globe is 5.1×10^{14} m^2. Thus, the volume of water constituting this instant deluge would be

$$.0254 \text{ m} \times 5.1 \times 10^{14} \text{ m}^2 = 13 \times 10^{12} \text{ m}^3$$

This is the water content of the reservoir, expressed in cubic meters rather than gallons.

Global precipitation in some combination of rain and snow constitutes the drain for this reservoir. How much does this amount to in the global and annual average? The data that we use to estimate this come from precipitation records at weather stations and ships at sea. For this calculation we shall take the answer to be the precise figure of 38.3 inches (0.972 m), remarking that the actual amount that falls is not so accurately known as this. With this figure we have the result that

$$0.972 \text{ m} \times 5.1 \times 10^{14} \text{ m}^2 = 496 \times 10^{12} \text{ m}^3$$

falls to earth in the course of a year.

When the reservoir content is divided by the rate of flow out the drain, we have the residence time.

$$\frac{13 \times 10^{12} \text{ m}^3}{496 \times 10^{12} \text{ m}^3 \text{ year}^{-1}} = 0.026 \text{ year} = 9.5 \text{ days}$$

The water that drains out of the global atmosphere is, of course, replaced by water vapor that is evaporated from the surface of the land and the ocean. This cycle of evaporation, temporary residence, and precipitation is the global-average hydrologic cycle. The surprisingly short residence time of about 10 days conveys the sense that the hydrologic cycle is very vigorous.

We began this chapter with a consideration of the role of weather in the global climate system and identified weather patterns like those seen in Figure 3.4 as the mechanisms responsible for the poleward transfer of heat that cancels the radiation imbalance shown in Figure 3.2. Now we can identity another global role for the weather patterns in Figure 3.4. The surface winds in these swirling patterns are evaporating water from the land and the ocean and are mixing this newly evaporated water throughout the global troposphere. The residence time of about 10 days that we established for the hydrologic cycle is so short that this mixing process is never completed. Water vapor soon condenses to form clouds, and this water falls back to earth. The distribution of water vapor over the globe at any instant is thus quite nonuniform. This is seen in photographs taken at infrared wavelengths sensitive to the presence of water vapor, such as the one at 6.7 μm in the lower lefthand panel of Figure 3.7.

Water vapor is unique among the greenhouse gases in having a residence time so short that its amount varies strongly with time and location. Determination of water vapor in the global atmosphere at any particular time thus requires a large number of observations. This task has traditionally been carried out by weather balloons that are equipped with humidity sensors and rise twice daily from weather stations distributed over the globe. A contrasting example is carbon dioxide. As we will see in a later chapter, it has a residence time of several years and is consequently well mixed throughout the global atmosphere. Then a measurement at a particular location such as that shown in Figure 1.1 provides a record of its global content.

REVIEW QUESTIONS

1. What is the function of weather in maintaining the balance between solar radiation absorbed by and infrared radiation emitted to space by the earth-atmosphere system?

2. At what latitude does the amount of infrared radiation emitted to space equal the amount of solar radiation absorbed when averaged over a period of a year or more?

3. Why are low clouds difficult to see in an infrared image?

4. What is window radiation? Why does an infrared sensor onboard a satellite use window radiation?

5. What can infrared satellite images tell us that visible images cannot?

6. How does chaos affect the accuracy of weather forecasts?

7. What are the steps involved in making a weather prediction based on the forecast of a numerical model?

8. What are some of the significant differences between a weather forecast and a climate simulation?

9. What is the definition of residence time?

10. How many meters (or inches) of water falls from the atmosphere in the global average each year? How many meters (or inches) of water are found in the form of water vapor in the atmosphere in the global and annual average?

INTERNET COMPANION

3.1 Why the Winds Blow

- The United States National Centers for Environmental Prediction (NCEP) and other agencies around the globe routinely monitor large-scale air motion in the troposphere. Information obtained from NCEP can be accessed from the Department of Meteorology Web server.

3.2 Satellite Cloud Imagery

- Current and archived satellite information is available from many sources. The Department of Meteorology Web server provides visible, infrared, and water vapor imagery routinely.

- The best source for many different forms of satellite imagery is the University of Wisconsin Space Science and Engineering Center. Short video clips of hurricanes and other special weather events are available.

3.3 The Prediction of Weather and Climate

- Current numerical weather predictions from NCEP are accessible from the Department of Meteorology Web server. The focus is centered, for the most part, on the prediction of weather over the western United States.

- Examples are readily available of output from one particular climate model, the Community Climate Model, Version 2, of the National Center for Atmospheric Research. The climate of this particular model is compared to the best available estimates of the observed climate.

3.4 The Hydrologic Cycle

- The NASA Marshall Space Flight Center provides in-depth information on the hydrologic cycle and estimation of precipitation around the globe.

CHAPTER 4

The Natural Variability of the Earth-Atmosphere System

Any global warming resulting from increasing atmospheric carbon dioxide would occur in the presence of the natural variability of climate that the earth-atmosphere system exhibits. In this chapter we look at two sources of natural climate variability that affect atmospheric temperature over periods of a few years in duration. These are volcanic eruptions and the temperature of the surface of the ocean. We will also look at the exchange of surface water and deep water of the ocean, which is a process that is suspected of producing variations of atmospheric temperature over periods around 1000 years long.

4.1 THE ROLE OF VOLCANOES IN THE EARTH-ATMOSPHERE SYSTEM

Large volcanic eruptions in tropical latitudes can affect global climate by injecting material into the global stratosphere. Before we consider the mechanism by which this can lead to temperature changes in the troposphere, we take up a more fundamental role that volcanoes have played: It is from volcanic emissions that much of the gas that constitutes the atmosphere has come.

The Composition and Origin of the Atmosphere

The earth is generally accepted to have been formed about 4.6 billion years ago without an atmosphere. The atmosphere slowly grew thereafter from the gases emitted by volcanoes. Volcanic emissions at present have roughly the following gaseous composition, which we take to be representative of volcanoes present on the early earth:

- 1% nitrogen
- 80% water vapor
- 12% carbon dioxide
- 7% sulfur dioxide and other gases

Consider now the early earth with its volcanoes. The gasses that they emitted began to accumulate to form an atmosphere. Since most of the emission was water vapor, the atmosphere (which then, as now, could hold only a little water vapor) soon reached the point at which it was saturated. After that time, the water vapor emitted by the volcanoes condensed to liquid water, and this began to form the global oceans.

The present gaseous composition of the atmosphere is:

- 78% nitrogen
- 21% oxygen
- less than 1% argon
- .4% water vapor
- .036% carbon dioxide
- tiny traces of other gases

How do we get from the composition characterizing volcanic emissions to the present-day composition of the atmosphere? First of all, the simple fact that the earth and the atmosphere are in contact with each other at the ocean surface permits molecules of atmospheric gas to enter the ocean. The total amount of each species of molecule (nitrogen, oxygen, etc.) that has found its way into the ocean in this way and resides there constitutes a *dissolved gas*. Thus, we find in the ocean dissolved nitrogen, dissolved oxygen, and so on. This is obviously what allows life in the ocean: The gills of fish are devices for extracting dissolved oxygen, and dissolved carbon dioxide feeds plant life in the ocean.

Ocean water at the present time contains the following amounts of dissolved gases:

- .9% nitrogen
- .5% oxygen
- 4.5% carbon dioxide

The interesting aspect of these numbers when compared with those for the atmosphere is that carbon dioxide evidently dissolves in ocean water much more readily than nitrogen and oxygen. (The reason is that nitrogen and oxygen do not react chemically with water, but carbon dioxide does.)

Although it dissolves readily in the oceans, the present amount of carbon dioxide in the ocean is an insignificant fraction of the total amount of carbon dioxide emitted by volcanoes since the formation of the earth. The early oceans soon became saturated with respect to dissolved carbon dioxide much as the ancient atmosphere became saturated with respect to water vapor. In ocean water saturated with dissolved carbon dioxide, the compound calcium carbonate forms and precipitates to the ocean floor. By this process, dissolved carbon dioxide has been steadily removed from the ocean and the carbon atoms deposited in sediments. This opened the way for more atmospheric carbon dioxide to dissolve in the ocean. The net result is that the carbon in the carbon dioxide emitted by the volcanoes over the ages has for the most part been returned to the earth. We will discuss this process in more detail in Chapter 6.

Now we can begin to see how the present atmospheric composition was achieved. Since nitrogen is an inert (nonreacting) gas in both the atmosphere and the ocean, the small amount of nitrogen emitted by volcanoes has accumulated in the atmosphere and in the ocean. In a similar manner, the tiny amount of the very inert gas argon emitted by volcanoes has accumulated to a detectable amount in the present atmosphere. In contrast, the history of carbon dioxide has not been continuous increase, but rather continuous removal.

Our last remaining puzzle is how the present abundance of oxygen was created in the atmosphere, since this gas is not found in volcanic emissions. Its presence can be explained by the process of photosynthesis, in which plant life in the oceans (and on land) consumes carbon dioxide and ejects oxygen. As we will see in Chapter 6, oxygen in the atmosphere owes its existence to photosynthesis in the ocean.

The Effects of Volcanoes on Our Present Climate

The eruption of a volcano always has an immediate impact on the weather in nearby regions. As the emitted water vapor cools and forms a cloud of water and ash high above and downwind of the volcano, the daytime temperatures drop noticeably and the nighttime temperatures increase. This is an example of the effect of high clouds on the total radiation balance. In addition to such local effects, the eruption of a major volcano has long been thought to be associated with changes in the global climate, most noticeably a decrease of surface temperature that lasts for a year or two after the eruption takes place.

First, what constitutes a major volcanic eruption? Volcanic eruptions of varying intensity take place around the globe every year. However, only a few fall into the major eruption category likely to have had an impact on the global climate. These include:

- Mount Pinatubo, Philippines, 1991
- El Chichón, Mexico, 1982
- Agung, Indonesia, 1963
- Santa Maria, Guatemala, 1902
- Krakatoa, Indonesia, 1883
- Tambora, Indonesia, 1815

Some scientists who are involved in research in this field might want to include a few more that took place during the past 200 years, but these six are the best documented. Notice that missing from the list is the most recent volcanic eruption in the continental United States, Mt. Saint Helens in Washington in May 1980. This eruption had an immediate impact on the weather downwind over the state of Washington and parts of Idaho, but it did not have any appreciable effect on the global climate.

The composition of the gaseous output from volcanoes was listed earlier in this chapter. However, there is another significant aspect of volcanic emissions: heat and the explosive energy of the eruption. The massive amounts of energy involved in major volcanic eruptions are enough to send the plume carrying water vapor, sulfur dioxide, and ash more than 20 to 30 km up into the atmosphere. As the plume is subsequently carried laterally by the prevailing winds, most of the heavier particles that constitute the ash and the water droplets falls out within a few hundred kilometers of the volcano.

We just mentioned that the volcanic plume can extend to heights in excess of 20 or 30 km. Recall from Figure 1.4 that the troposphere constitutes the lowest 10 km, while the stratosphere extends upward another 30 km. In other words, volcanic plumes in major eruptions penetrate far into the stratosphere. The stratosphere is characterized by stable conditions; once a lightweight particle reaches this elevation, it is likely to remain suspended in the stratosphere for a long time. Put another way, the residence time of the particles in the stratosphere is much longer than that of similar particles in the troposphere.

A significant aspect of the list of major volcanoes is that they all are situated well equatorward of 30° latitude. The reason that this is significant is that when a volcanic plume enters the tropical stratosphere, it can spread out laterally and poleward into both hemispheres. For this reason the major volcanic eruptions for climate purposes are those that take place in the tropics, and the closer to the equator, the better. The area covered by a plume from a similar volcano outside the tropics is limited by the prevailing tendency of the atmospheric circulation in the stratosphere to carry material poleward from the entry location, and this leaves the tropics and the opposite hemisphere free of the material in the plume.

What is the mechanism by which volcanic eruptions affect the global climate of the earth-atmosphere system? This question is still under investigation, but there is one very strong candidate. Volcanoes emit a significant amount of sulfur dioxide, which mixes with the water vapor in the plume and in the stratosphere to form tiny droplets and particles of sulfuric acid called sulfate aerosol. This aerosol remains suspended in the stratosphere and is spread laterally by the stratospheric circulation. If the volcano is close to the equator, the spreading will be global, as we have mentioned. The amount of sulfur dioxide emitted into the atmosphere during volcanic eruptions can vary widely: Mt. Pinatubo is estimated to have emitted over 20 times more sulfur dioxide than Mt. Saint Helens.

One consequence of the additional sulfate aerosol suspended in the stratosphere is brilliant red and orange sunsets that last from several months after a major eruption until a couple of years later. The delay of a few months is the time that it takes to spread the aerosol from the tropics to the mid-latitudes. Most recently, the eruption of Mt. Pinatubo led to vivid sunsets during 1991 and 1992.

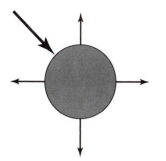

Figure 4.1 Incoming solar radiation (heavy arrow) is shown being intercepted by an aerosol particle in the atmosphere. The solar radiation is scattered (thin arrows) in all directions.

We can infer from this that the residence time of sulfate aerosol in the stratosphere is roughly a couple of years.

How do these aerosol particles act to produce the sunset colors? They promote *scattering*, a process by which sunlight shining on a small solid particle that is suspended in the atmosphere is reflected in all directions as shown schematically in Figure 4.1. Scattering from molecules or from small particles depends on the wavelength of the solar radiation, and short wavelengths (that is, the blue end of the visible spectrum) are preferentially scattered at the expense of the long (red) wavelengths. Scattering from the molecules that constitute the atmosphere is responsible for the normally blue color of the clear sky during most of the day. However, at sunrise and sunset, the sunlight travels farther through a thicker portion of the atmosphere, as shown in Figure 4.2, so that the blue light is scattered out of the solar beam and what remains to be seen is orange and red light. The extra contribution to scattering by enhanced sulfate aerosol that lingers in the stratosphere following a major eruption contributes to especially vivid sunsets.

Now we return to consideration of the climatic impact of major volcanic

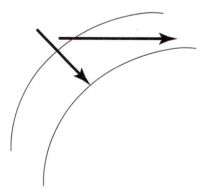

Figure 4.2 The distance that light travels through the atmosphere is shorter at noon when the sun is nearly overhead than it is at sunrise or sunset.

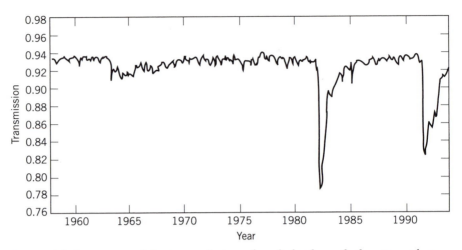

Figure 4.3 A measure of the transmission of sunlight through the atmosphere determined at the Mauna Loa Observatory, Hawaii, over the past 35 years.

eruptions. It has been hypothesized that the additional sulfate aerosol in the stratosphere reflects a significant amount of visible solar radiation back to space. As the amount of aerosol in the stratosphere increases after an eruption, the amount of sunlight reaching the earth's surface diminishes and the global-average surface temperature should decrease. Figure 4.3 shows the reduction of the transmission of sunlight through the atmosphere after the eruptions of Agung, Chichón, and Pinatubo. The reduction after the Agung eruption is quite small, but that after the Chichón and Pinatubo eruptions is quite noticeable. The reduced transmission of sunlight after the Chichón and Pinatubo eruptions indicates that the enhanced aerosol affects the atmosphere for a year or two.

The link between volcanic eruptions and climate variations is becoming more firmly established. Monitoring the climate system by satellite before, during, and after the eruption of Mt. Pinatubo has provided an improved understanding of the spread of volcanic aerosols around the globe. However, the impact of volcanic eruptions on the global-average surface temperature is small; the record suggests that it may drop a couple of tenths of a degree Centigrade after a major eruption for a year or two. Statistical analysis of the effects of major volcanic eruptions is difficult because so few have occurred. Some evidence of the link between global climate and major volcanic eruptions is provided in Figure 4.4. This shows the temperature record of Figure 1.5 with the respective times of eruptions of Krakatoa, Santa Maria, Agung, El Chichón, and Pinatubo indicated by arrows. Dips are evident in the temperature record after the Krakatoa eruption in 1883 and the Agung eruption in 1963. However, there is no such dip after the 1982 eruption of El Chichón. The much warmer than normal conditions during the 1980s are seen in Figure 4.3 to have diminished after the time of the Mt. Pinatubo volcanic eruption.

We have presented here a discussion of major volcanic eruptions as an example of a physical phenomenon that very likely affects global climate, yet for

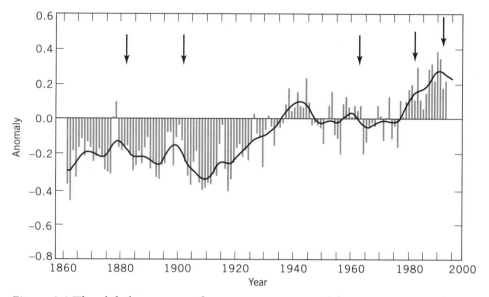

Figure 4.4 The global-average surface temperature record from Figure 1.5 with the years of major volcanic eruptions indicated by arrows.

which it has proved difficult to determine exactly how large the effect is. There is no doubt that the amount of solar radiation transmitted through the stratosphere is reduced during and after a major volcanic eruption. Thus, we have a strong causal link: Volcanoes reduce the amount of solar radiation reaching the surface. However, we cannot demonstrate that this is always associated with a significant decrease in the global-average surface temperature.

4.2 AIR-SEA INTERACTIONS

The concept of *heat transfer* comes up in many different settings in the atmosphere. Solar radiation is transfer of heat from the sun to the earth-atmosphere system, and infrared radiation is transfer of heat from the earth-atmosphere system to space. We have also seen that poleward heat transfer in the atmosphere and the ocean are what, in the annual average, cancels the latitudinal imbalance of radiation. We now turn to examine yet another example of heat transfer: that between the atmosphere and the oceans. Exchange of infrared radiation between the atmosphere and the surface of the ocean is one such mechanism of heat transfer. Another is evaporation from the sea surface and subsequent condensation of this water vapor in the atmosphere.

This section focuses on a third mechanism of heat transfer between the atmosphere and the ocean: that which occurs when cold air blows over warm water or when warm air blows over cold water. We will not go into the details of the physical process that produces a vertical transfer of heat in these situations. An appeal to experience or simple intuition is sufficient. When cold air is moving across the surface of a warm body of water, the air gets warmer as it goes along.

This constitutes a transfer of heat from the water to the air. Similarly, when warm air is moving across the surface of a cold body of water, the air gets cooler as it goes along, and this is the result of heat being transferred from the air to the water.

The Structure of the Oceans

The ocean covers about 70 percent of the surface of the earth and has an average depth of 3700 m. The ocean can be divided into two layers: the *mixed layer*, occupying on average the top 100 m (330 ft); and the deep ocean, which is everything below. One reason for this division into two layers is that sunlight does not penetrate more than about 100 m into the ocean. As a result, the mixed layer of the ocean is warmer than the deep ocean (Figure 4.5). Another feature that characterizes the mixed layer is that it is well stirred (mixed) by the action of the winds blowing across the ocean surface. The mixed layer can be viewed as the part of the ocean that is responsive to changes in atmospheric conditions. The net effect of this agitation by the atmosphere is that the mixed layer of the ocean tends to be uniformly warm throughout its 100-m depth. Because warm water is also light water, we can say literally that the mixed layer floats on top of the deep ocean. The rest of the ocean below is dark and cold. Its only communication with the mixed layer (and the atmosphere above) is by the processes of upwelling and downwelling that we now consider.

The simple sketch in Figure 4.5 of the vertical structure of the ocean applies everywhere over the earth except in relatively small but important regions in high latitudes. There, the surface of the ocean is in contact with very cold air during much of the year. Heat transfer from the ocean to the cold air in those regions renders the near-surface water so cold that it sinks (downwells) into the deep ocean. As illustrated in Figure 4.6, there is then no mixed layer in these regions, and the ocean there is just one continuous body of cold, near-surface water sinking into the depths.

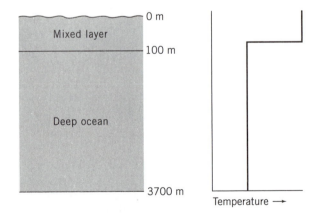

Figure 4.5 A simple schematic of the layers of the ocean and an idealized profile of temperature with depth.

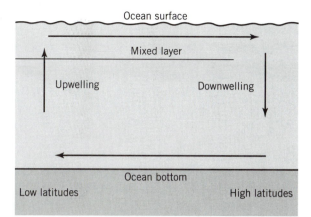

Figure 4.6 Schematic of the thermohaline circulation. The mixed layer is absent in high-latitudes, where the atmosphere removes heat from the ocean surface and the resulting cold water sinks to great depths.

This water that sinks must come back up somewhere. The rising of water from the ocean depths is called *upwelling*. The downwelling occurs at high latitudes, and then the water moves equatorward in the deep parts of the oceans and is upwelled again at lower latitudes. The downwelling water in turn is replaced by water that has to move poleward from lower latitudes. What we then have, as illustrated in Figure 4.6, is a circulation cell with the same directional sense as the one for the atmosphere that was shown in Figure 3.3. What this ocean circulation is doing is transporting cold water south (deep in the ocean) and warm water north. Like its atmospheric counterpart, it thus transports heat poleward. For reasons that we go into later, it is called the *thermohaline circulation*.

Apart from the thermohaline circulation, there is also a global pattern of *wind-driven* currents that are confined to the mixed layer. These currents in the near-surface water are actually responsible for most of the heat that the ocean transports poleward. An example that is probably familiar is the Gulf Stream, the warm current that flows northward along the east coast of the United States. Wind-driven currents in the mixed layer also give rise to *coastal upwelling*. This phenomenon is responsible for the relatively cold surface water along the west coast of the United States.

We return to our consideration of the thermohaline circulation by remarking that its actual configuration is more complicated than that illustrated in Figure 4.6. It is quite literally a global circulation, and its pattern is illustrated in Figure 4.7. As seen there, the downwelling occurs in the North Atlantic region. That is because the cold (and, hence, heavy) surface water found there is made even heavier by virtue of being just about the most salty surface water anywhere in the ocean. The key words "cold" and "salty" are embodied in the term "thermohaline."

As shown in Figure 4.7, the water sinks in the North Atlantic and flows southward at depth into the Antarctic Ocean, where it turns eastward. At that point it is joined by some water (not shown in this diagram) that is sinking in a

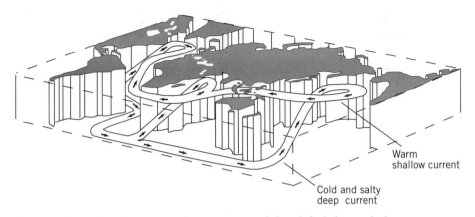

Figure 4.7 A three-dimensional perspective of the global thermohaline circulation. Continents are shown in black.

region near Antarctica. This cold, salty water is the by-product of ice formation around the edge of Antarctica. As we follow the stream of deep water eastward, we see that some of it upwells in low latitudes of the Indian Ocean and the rest upwells in low latitudes of the Pacific Ocean. The water then makes its way in the mixed layer back through the Indonesian region, across the Indian Ocean, around the tip of Africa and back into the North Atlantic. Not shown in this diagram is a feeder branch of this surface circulation that brings surface water back to the vicinity of Antarctica to replace the surface water sinking there.

By this elaborate means the deep ocean maintains global contact with the mixed layer and the atmosphere. This global circulation is very slow. It has been estimated that water that enters the deep ocean by downwelling in the North Atlantic takes roughly 1000 years to reach the surface again in the Pacific. This long residence time of water in the deep ocean stands in marked contrast to the residence time of water in the atmosphere, which was seen in Chapter 3 to be about 10 days.

The Heat Capacity of the Atmosphere and Ocean

Consider now a local region of the ocean and suppose that what is occurring at the moment of interest is a transfer of heat from the atmosphere to the ocean. Near the surface, the atmosphere cools and the water warms. But, for a given transfer of heat, the temperature increase of the water will in general be much smaller than the temperature decrease of the air. This is due to two things: The *specific heat* and the *density* of water are each greater than the specific heat and the density of air. Specific heat is defined as the heat that must be transferred to raise by 1°C the temperature of 1 kg of a substance. The specific heat of water is 4.186 times that of air. A factor of even greater significance is that the density of water is 10^3 kg m^{-3}, while that of air is 1 kg m^{-3}.

In our example of a moment of local exchange of heat between atmosphere and ocean, the temperature change is confined to a region near the ocean surface.

When such heat exchange is viewed in the context of global or even regional climate, the vertical domain affected by the heat transfer is much greater. This domain extends through much of the troposphere and down through the whole of the mixed layer of the ocean. The deep ocean is not affected by such heat exchange, except in the special high latitude regions noted in the preceding section. Now consider a column of 1 m² cross section extending from the base of the mixed layer of the ocean upward to the top of the troposphere. Since the mixed layer is 100 m deep, the volume of water in the column is 100 m³. Given the density of water as 10^3 kg m^{-3}, this water weighs 10^5 kg. Since the troposphere is 10 km deep, the volume of air in the column is 10^4 m³. This air weighs 10^4 kg.

Now let an arbitrary amount of heat be transferred from the air in the column to the water in the column. The increase of temperature of the water will be smaller by a factor of 41.86 than the decrease of temperature of the air. In this sense, the mixed layer of the global ocean serves as a considerable reservoir of heat in the earth-atmosphere system. It can return this heat to the atmosphere when and where colder air flows over it. Or it may transfer this heat to the atmosphere by infrared radiation or in the form of latent heat in water vapor evaporated from its surface. To complete the list of things that transfer heat to or from the ocean, we must not forget solar radiation. The global ocean extends over 70 percent of the surface. All the solar radiation absorbed at the ocean surface represents a transfer of heat directly from the sun to the mixed layer.

El Niño

A strong form of regional air-sea interaction is found in the equatorial Pacific Ocean. Let's begin with Figure 4.8, showing the normal distribution of sea-surface temperature across the tropical Pacific Ocean during the month of October. This represents the climate state of the ocean surface during that month. The warmest

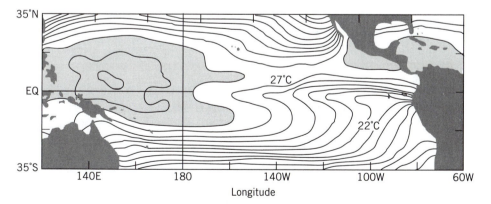

Figure 4.8 Sea-surface temperature across the tropical Pacific Ocean during October based on many years of observations. Areas in which the temperature exceeds 28°C are shaded. The temperature field is contoured at an interval of 1°C.

region of the ocean is found in the western Pacific (mostly to the west of the International Date Line, 180° longitude); over that vast area, the sea surface temperature exceeds 28°C. As we look farther east along the equator, the ocean surface becomes colder and colder, until we find the temperature along the coast of South America to be less than 23°C. Temperatures are even colder farther south along the coast of South America.

This climate state of the ocean is determined primarily by the stress placed on the ocean surface by the winds in the atmosphere immediately above the surface. The climate state of the surface wind field is characterized by easterly winds (blowing from east to west) along the equator that help to drive ocean currents in the same direction. The cold waters found along the South American coast are then carried (advected) westward out along the equator. The easterly winds weaken considerably as they approach the International Date Line, so much so that the waters of the western Pacific remain warm year-round.

Every few years, the sea surface temperature in the equatorial Pacific undergoes significant fluctuations. In some years, the ocean surface warms up; the largest warming during the past 100 years occurred in 1982. The top panel of Figure 4.9 shows the distribution of sea-surface temperature during November of that year. Notice that temperatures higher than 28°C extend much farther east (to around

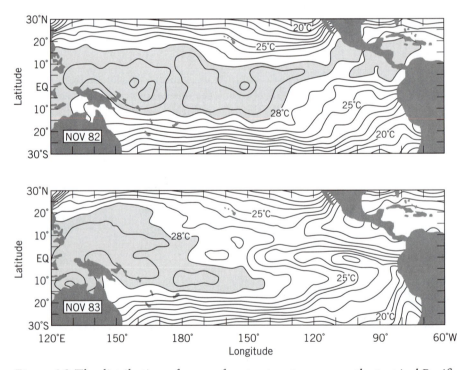

Figure 4.9 The distribution of sea surface temperature across the tropical Pacific Ocean during El Niño conditions (top panel) and La Niña conditions (bottom panel). Areas in which temperature exceed 28°C are shaded. The temperature field is contoured at an interval of 1°C.

130°W along the equator) compared to the climate state (Figure 4.8). While it may be difficult to assess by comparing the top panel of Figure 4.9 and Figure 4.8, the largest changes in sea-surface temperature relative to the climate state occurred near the coast of South America, where the temperatures increased by as much as 3°C. One year later (the bottom panel of Fig. 4.9), the distribution of sea-surface temperature is quite different. It looks like an enhanced version of the climate state, with a pronounced tongue of cold water extending westward from South America toward the International Date Line.

The basin-wide changes in sea-surface temperature exhibited in Figure 4.9 went largely undetected until the past 40 years. However, sailors and residents of Peru and Ecuador during the late 1800s were aware of significant variations every few years in the currents and weather along the west coast of South America. Starting around Christmas, currents carrying warm water southward along the coast were observed. Torrential rainfall along the normally arid (desert) coastal plain often began as well around the same time of year. Coastal areas that were unable to support even the hardiest vegetation during most years became covered by extensive fields of grass and other plants. The occurrence of these events shortly after Christmas led them to be referred to collectively as El Niño in reference to the Christ child.

During the early part of the twentieth century, a negative aspect of El Niño became apparent: widespread disappearance of some fish species and severe mortality of bird populations. The fisheries in this region are remarkably prolific as a result of strong coastal upwelling. Nutrients are carried upward from the deep ocean into the mixed layer where, in the presence of sunlight, tiny plant life flourishes. This, in turn, provides the basis for a vigorous food chain. During El Niño, fish species dependent upon this food chain suffer, while a few species that can tolerate warmer, nutrient-deficient water survive.

The plight of the residents along the coast of Peru is interesting, but such local effects have little relevance to the issue of global climate change. During the late 1950s and early 1960s scientists began to realize that the coastal El Niño phenomenon was associated with changes in the ocean across the entire Pacific as shown in the top panel of Figure 4.9. Hence, the appearance of abnormally warm water across the Pacific began to be called El Niño as well. Scientists also have a tendency to corrupt the meaning of terms for convenience. Years in which the temperatures are abnormally cold along the coast of Peru and the equatorial Pacific are now called La Niña years. The conditions found in November 1983 (lower panel of Fig. 4.9) can be considered to be representative of La Niña. The most recent El Niño episodes have occurred in 1987, 1992, and 1994 while the most recent La Niña events occurred in 1988 and 1995.

The temperature changes associated with El Niño and La Niña cover such a broad extent of the equatorial Pacific Ocean that they directly affect the surface temperature of the entire tropical strip as shown in Figure 4.10. There are periods (El Niño years) when the temperature in the tropical strip is much higher than usual and periods (La Niña years) when it is much lower than usual. The magnitudes of the changes in temperature here from year to year are small, on the order of 0.5–1°C. However, these temperature swings have major effects on the climate of the earth-atmosphere system.

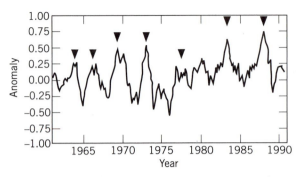

Figure 4.10 Month-to-month variations in surface temperature in the tropical strip (20°N–20°S) from 1961–1989. El Niño episodes are indicated by arrows.

The El Niño/Southern Oscillation (ENSO) Phenomenon

Of even greater importance than the direct warming or cooling of the sea surface is the indirect effect of these changes on the atmosphere above and the ways in which the ocean and atmosphere interact. El Niño is initiated when the easterly surface winds across the Pacific weaken and La Niña begins when the winds strengthen. This represents a coupling between the ocean and atmosphere that has profound effects on the earth-atmosphere climate system. The coupled interactions between the atmosphere and ocean in the tropical Pacific are now called the El Niño/Southern Oscillation phenomenon, which we will abbreviate as the ENSO phenomenon. The term Southern Oscillation refers to variations in the pressure field of the atmosphere that are associated with the weakening and subsequent strengthening of the winds during the El Niño/La Niña cycle. ENSO can be thought of as describing the extremes in the climate of the equatorial strip of the earth embracing the years of both abnormally cold and abnormally warm ocean waters in the equatorial Pacific.

Why is the ENSO phenomenon so important? First, it provides a glimpse of the complex workings of the earth-atmosphere system. Because this phenomenon occurs every few years, we can see how the components of the climate system vary without having to wait decades, as with the global warming problem. Second, the tropical Pacific Ocean covers a significant portion of the globe; the changes that take place there constitute a significant part of the global average. The abnormal warming since the late 1970s, seen in the global surface temperature record (Fig. 1.5), can be explained in part by unusually frequent and pronounced El Niño episodes without as many strong La Niña periods (see Fig. 4.10). Third, the ENSO phenomenon has many of the characteristics of a chaotic system, that is, the maddening mixture of ordered and random behavior. Considerable effort has been placed on developing models to predict the occurrence of ENSO events. These same improvements can be used in climate models to improve their simulations of the effects of increased carbon dioxide emissions. Finally, the ENSO phenomenon affects the weather around the globe, including the United States. However, in keeping with the chaotic nature of the atmosphere, there appears to be no single weather pattern over the United States associated with El Niño or La Niña.

The earth-atmosphere climate system is not like a flask in a chemistry laboratory where a complex chemical reaction can be studied under controlled conditions. Rather, our climate system is constantly undergoing changes on all time scales. One difficulty in understanding the potential climate effects of increasing carbon dioxide is that the present situation has not happened before; therefore, we don't know what to expect. Scientists in such situations often look for other behavior that has occurred many times before and that can be used as a reasonable proxy for what they really want to study. The ENSO phenomenon can, in some respects, be viewed as a proxy for longer-term climate variations. ENSO episodes occur every few years, and the last few events have been studied in great detail.

We mentioned briefly in the last subsection that the equatorial Pacific Ocean responds to changes in the strength of the easterly winds along the equator. In other words, we have viewed the surface of the ocean (and more accurately, the entire mixed layer) as being forced by the atmosphere. However, the greater heat capacity of the oceans' mixed layer compared to the atmosphere allows the mixed layer to force the atmosphere for long periods.

The primary means for the ocean to force the atmosphere is through changes in sea-surface temperature, especially in the equatorial Pacific Ocean. What is the significance to the atmosphere of a 1° or 2°C increase in sea-surface temperature in the equatorial Pacific such as seen in Figure 4.9? First, the amount of water that evaporates from the ocean depends nonlinearly on the sea-surface temperature. As the temperature increases slightly, the amount of water that can be evaporated increases sharply as shown schematically in Figure 4.11.

The increase in the rate of evaporation from the Pacific Ocean during El Niño years and the decrease in evaporation during La Niña years is not enough to explain how the ocean forces the atmosphere in this region. The schematic in Figure 4.12 shows the way that a small increase in sea-surface temperature can lead to a large effect on the atmosphere. As a patch of the equatorial Pacific becomes warm relative to its surroundings, the amount of water that evaporates into the atmosphere increases and the air above warms as well. As described in section 3.1, warm areas in the atmosphere tend to be associated with lower pressures. This leads to a circulation in the atmosphere where the air near the surface tends to blow toward the warm air, in order to attempt to fill that trough of low pressure.

As the air converges toward the region of warm air, it is forced aloft and cools

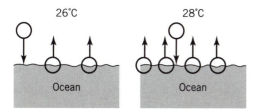

Figure 4.11 As the sea surface temperature increases from 26°C to 28°C, the number of molecules of water that evaporate from the ocean surface increases markedly.

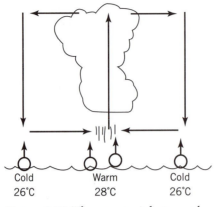

Figure 4.12 The equatorial atmosphere can respond sharply to regional changes in sea-surface temperature as discussed in the text.

as it is lifted. Since the air is full of water vapor, some of it will condense to form clouds, and eventually, if there is enough lifting, rain will occur. Remember that evaporation of water from the ocean extracts heat from it, while condensation of water vapor releases heat in the atmosphere. In this manner, the formation of the clouds and rain leads to additional heating of the atmosphere and the process continues until the ocean cools relative to its surroundings or the atmosphere stabilizes the circulation in this region so that additional clouds are unlikely to

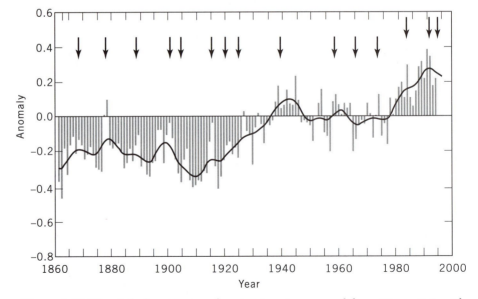

Figure 4.13 The global-average surface temperature record from Figure 1.5 with major ENSO episodes indicated by arrows.

form. Finally, note in the diagram that the air returns aloft and subsides over the colder waters. Thus, during La Niña, the atmosphere over the central equatorial Pacific tends to be less cloudy with less rainfall, since the ocean temperatures are colder than normal.

We will end this subsection with a brief examination of the effects of ENSO episodes on the global-average surface temperature field. Earlier (Figure 4.10) we saw that the ENSO phenomenon controls the surface temperature in the tropical strip, but does it affect the global temperature as strongly and over the full length of the record? Figure 4.13 repeats the global-average temperature record with major El Niño episodes indicated by arrows. As with the major volcanic eruptions, the occurrence of El Niño episodes cannot explain all the variations in the global climate record. However, it is possible to see that several recent periods of warming in the equatorial Pacific are related to warmer than normal conditions in the global average. In addition, El Niño years before 1930 tended to occur at times when the global-average temperature was higher relative to contemporary years.

R E V I E W   Q U E S T I O N S

1. How did the ancient atmosphere and the oceans form from volcanic emissions?

2. What are the processes that have made the chemical composition of the atmosphere so different from the composition of volcanic emissions from which it originated?

3. Why are sunsets redder than usual several months after major volcanic eruptions?

4. Why is more solar energy absorbed by the oceans than by soils?

5. What are the differences between the mixed layer of the ocean and the deep ocean that lies beneath it? Why is there no mixed layer in the high latitude oceans?

6. In what part of the world does surface water sink into the deep ocean and in what part of the world does deep ocean upwell to the surface? What causes surface water to sink?

7. What are the significant differences between El Niño and La Niña episodes?

8. Of what value is research on the ENSO phenomenon with respect to the global warming issue?

INTERNET COMPANION

4.1   The Role of Volcanoes

• The NASA Goddard Space Flight Center provides current information on recent eruptions in cooperation with many other agencies and scientists.

- An overview of the role of volcanoes in climate variations has been made available by NASA. Measurements of sulfur dioxide from satellite-borne instrumentation (the NASA Total Ozone Mapping Spectrometer) indicate the climate impact of different volcanic eruptions.

4.2 Air-Sea Interactions

- The Smithsonian Ocean Planet exhibit is a fascinating interactive look at many aspects of the world's oceans. The Climate Prediction Center of the National Centers for Environmental Prediction provides current information on the state of the tropical atmosphere and ocean and the ENSO phenomenon.

- The El Niño theme page developed by the NOAA Pacific Marine Environmental Laboratory provides access to information and data related to El Niño.

- The NOAA Report to the Nation on Our Changing Planet entitled "El Niño and Climate Prediction" provides useful background on the ENSO phenomenon.

CHAPTER 5

Long-Term Climate Variations

The preceding chapter examined two sources of natural year-to-year variation of global climate. These sources of short-term climate variation were eruption of a major volcano located in the tropics and changing patterns of sea-surface temperature, respectively. We saw that their influence on the global-average surface temperature amounts to no more than a few tenths of a degree Centigrade. These small fluctuations are superimposed on longer term fluctuations that characterize the global-average surface temperature record (Fig. 1.5). To review, an increase of about 0.5°C has been observed over the past century. As we look backward over a much longer interval of time, we see changes in global-average surface temperature possibly as large as 5°C. These long-term variations are the subject of this chapter.

5.1 PALEOCLIMATOLOGY

The Paleoclimate Record

The network of weather-observing stations that has provided the measurements that go into the global-average surface temperature record (Fig. 1.5) built up slowly from its inception some 135 years ago. This was preceded by a period of widely scattered temperature measurements. For still earlier times, temperature has to be inferred from other data, whose nature we will examine later. Here we state the important point that any global-average surface temperature record before about 135 years ago is an estimation. One such estimated record is shown in Fig. 5.1.

The upper panel of this figure shows an estimate of global-average surface temperature variations over the past million years. Notice that over this span of time, the temperature has rarely been higher than the present value. The particularly cold periods in the record are glacial episodes, more commonly referred to as *ice ages*, in which a significant fraction of the land area poleward of 50°N was covered by ice. The intervening warmer periods are interglacial episodes in which this polar ice cap of the Northern Hemisphere retreated. The ice ages occur at irregular intervals usually spaced about 100,000 years apart. The last ice age began abruptly around 120,000 years ago and was preceded by one of the warmest interglacial periods of the past million years. This interglacial period is called the Eemian interglacial. The last ice age reached its peak about 18,000 years ago and then ended abruptly 12,000 years ago. We have been in an interglacial period, referred to as the Holocene, for the last 10,000 years.

The middle panel of Figure 5.1 summarizes the changes in temperature over the past 11,000 years. The conditions at the left edge of the figure reflect the final stages of the last ice age. Roughly 4000 to 6000 years ago, the earth-atmosphere climate system was characterized by warmer conditions than those at the present time. This period is called the Holocene maximum, and it represents the warmest period to date during the present interglacial.

The bottom panel in Figure 5.1 displays the variations in temperature during the past thousand years. Between A.D. 1000 to 1300, the globally averaged temperature was a few tenths of a degree warmer than that observed now. After that there was a cooling trend, which has been called the Little Ice Age. Mountain glaciers in many regions of the world expanded significantly during this period. The warmer conditions observed since the middle part of the nineteenth century are what we looked at in Figure 1.5.

In Chapter 2 we referred to the intensity of solar radiation at the distance from the sun where the earth is situated as the solar constant and stated its value to be 1368 W m^{-2}. Figure 5.2 shows that the solar constant has varied over a range of 1 W m^{-2} over the past hundred years, with the most pronounced variations occurring in the past 40 years. These semi-regular cycles in the solar constant have a period of roughly 11 years. A similar 11-year cycle has also been observed in the number of sunspots (dark blotches) on the surface of the sun. Comparison of sunspot number and satellite measurements of the solar constant indicate that the solar constant is higher during periods when the number of sunspots is higher. Attempts to relate variations in the number of sunspots to changes in weather and climate during the present century have met with little success. On the other hand, the Little Ice Age at its peak during the 1600s has been linked to the occurrence of a reduced number of sunspots.

What relevance does the behavior of temperature over thousand-year to million-year time periods have for the global warming problem? Most important, it shows that the earth-atmosphere system has undergone large temperature swings within a fairly well-defined temperature range. During the past million years, the earth's surface temperature has varied over a range of about 5°C. This range is

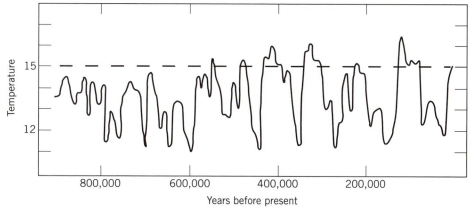

(a)

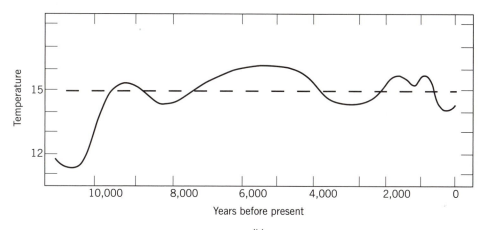

(b)

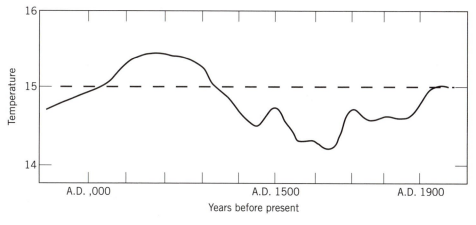

(c)

Figure 5.1 Estimated global-average surface temperature variations during: (top) past million years, (middle) past ten thousand years, and (bottom) past thousand years. The dashed line across each panel of the figure indicates the present value of 15°C.

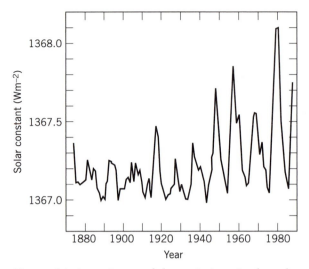

Figure 5.2 An estimate of the variations in the solar constant over the past 100 years. The values along the left margin refer to the solar constant in W m⁻².

large compared to the 0.5°C warming that has occurred over the past 100 years, but is not all that much larger than the global warming of 2°C that is predicted to occur in the next century. (We will look at the basis of this prediction in Chapter 8.)

We showed in Figure. 1.1 the observed increase in carbon dioxide over the past 40 years. We show now in Figure 5.3 estimates of its variation over the past 160,000 years and over the past 250 years, respectively. These long-term records of carbon dioxide were determined from air bubbles trapped in the deep ice of Antarctica. During the present interglacial and the previous interglacial (around 120,000 years ago), the amount of carbon dioxide is seen to be higher than during the intervening ice age. Carbon dioxide in the atmosphere and temperature thus seem to be related. What has not been determined from such paleoclimate records is whether the periods of high carbon dioxide cause the high temperature or vice versa. The bottom panel of Figure 5.3 shows that, prior to the industrial age, the concentration of carbon dioxide was about 280 ppm, and it has risen steadily since then.

Paleoclimate Indicators

The global-average surface temperature record in Figure 1.5 is based on observations of temperature over the past 135 years. Temperature measurements extend back to the late seventeenth century for a couple of locations in Europe. If we wish to estimate the global-average surface temperature prior to 1860, we must rely on other sources of information.

A variety of methods have been developed to extract so-called proxy data to

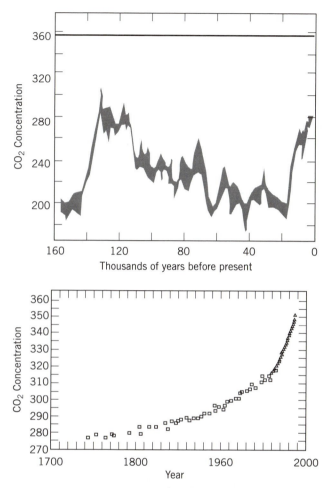

Figure 5.3 Concentration of carbon dioxide in ppm over the past 160,000 years (upper panel) and the past 250 years (lower panel). The line in the upper panel indicates the present concentration of carbon dioxide. The heavy shading in the upper panel indicates the range of values of the data.

determine long-term variation in temperature. These methods fall into two main categories:

- Faunal and floral techniques. These methods rely on natural recording systems such as the annual growth of trees and the rate of growth of coral reefs in the oceans.
- Sedimentological and stratigraphic techniques. These methods rely on analyses of the layering structure and chemical composition of soils and ice to infer climate information. Measurement of the relative amounts of different isotopes of oxygen in the sediments at the bottom of the ocean or in the polar ice caps is an example.

There is no one method that provides all the information required to determine past climate. The record in Figure 5.1 is based on a variety of paleoclimate data, all of which are subject to considerable uncertainty. Each indicator reflects a complex, integrated response to a number of climate variables. For example, the rate of growth of trees as measured by the width of the annual growth responds not only to the temperature but also to the amount of moisture available.

These proxy paleoclimate indicators are the only means available to document the history of the climate of the earth-atmosphere system. Another approach is to infer what the climate might have been in previous epochs by means of simulation with a mathematical climate model of the atmosphere as described in Chapter 3. Instead of using the present conditions to begin the model simulations (such as the current locations of the polar ice caps and coastlines), the conditions appropriate to an earlier period are used. These types of simulations provide detailed information on how past climates might have been constituted. However, there is limited paleoclimate data to verify whether they are accurately simulating the atmosphere of that era.

5.2 THE ICE AGES

A Brief Description

An ice age is characterized by the accumulation of ice on land in the high latitudes of the Northern Hemisphere. We are presently in an interglacial period, and accumulated ice is restricted largely to Greenland. This situation is contrasted in Figure 5.4 with that at the height of the last ice age (18,000 years ago), when a much larger area was covered by ice sheets. The enhanced albedo of the earth-atmosphere system resulting from this coverage was responsible for the cooler global climate experienced at the time.

The ice sheet at the height of the last ice age was not only extensive but also deep. It had a typical thickness of some 3 km (10,000 ft). This characteristic enables us to see why it is the Northern Hemisphere that is involved in the development of ice ages. In our present interglacial period Antarctica is covered by an ice sheet 3 km in thickness. This ice sheet, which extends all the way to the ocean encircling the continent, is in a steady state in which the amount of ice deposited as snow annually is removed in icebergs that break off into the ocean. Because of this balance, the Antarctic ice sheet was not very much thicker in the past ice age than it is now. It was also not any larger. Then, as now, it covered the continent and could not enlarge because a continental base is required to support the weight of a thick ice sheet.

The continental land masses in the Northern Hemisphere extend into very high latitudes and serve as platforms upon which ice sheets can grow equatorward when conditions are favorable. The mechanism for this growth is accumulation of winter snowfall that does not melt during summer. This is a key element in the currently accepted theory of how ice ages form: What is required for an ice age is a long epoch of cool summers in the Northern Hemisphere. What is also involved in the slow and steady growth of the area covered by ice sheets as development of the ice age proceeds is the *ice-albedo feedback* process illustrated in Figure 5.5.

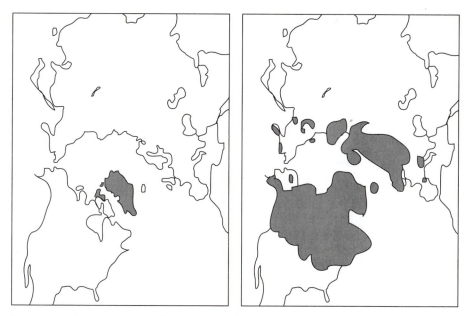

Figure 5.4 Permanent ice sheets now (left panel) cover only a small fraction of the land mass that they covered during the height of the last ice age (right panel).

The snow that falls in the Northern Hemisphere as the ice age develops forms in the atmosphere from water vapor that is evaporated from the ocean. It follows that, as ice accumulates on land during the long development of an ice age, global sea level falls. Otherwise stated, the water that gets locked up in ice sheets on land comes out of the global ocean. We will in the next section review evidence that shows that sea level at the height of the last ice age was some 120 m (400 ft) lower than it is at the present.

Determining the Ice-Age Record

We can tell something about an ice age from the effect of ice sheets on land areas of the Northern Hemisphere. As the ice sheets grew to great thickness, they pressed down the supporting land. This land is still slowly rebounding from the last ice

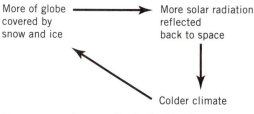

Figure 5.5 The ice-albedo feedback loop.

age, which ended 12,000 years ago. From the measured rate of rebound, the thickness of the ice sheet can be inferred. The maximum southward extent of the ice sheet not only in the last ice age but also in several preceding ice ages is marked by accumulations of boulders.

The single most important record of ice ages is that of past sea level. The water that is in the ice on land evaporated from the ocean. We earlier mentioned that sea level was 120 m (400 ft) lower at the peak of the last ice age than it is now. A reduction in the amount of water it contains is accompanied by a subtle change in the chemical composition of the global ocean. This progressive change in chemical composition of ocean water as an ice age develops is recorded in sediments that are always piling up on the ocean bottom. To recover this record the crew of a research ship drops a hollow cylinder that sinks into the sediment on the ocean bottom and is then winched back on board. This yields a *sediment core* that constitutes a sample of sediments that are progressively older the farther down the core they are located.

The specific chemical analysis problem is to determine at various places along this core of deep-sea mud the amount of the relatively rare isotope of oxygen known as oxygen eighteen (denoted ^{18}O) that is present in the compound calcium carbonate that constitutes the mud. Appendix A.4 briefly describes what an isotope is. As we explain next, the amount of the isotope ^{18}O relative to the amount of the isotope ^{16}O in ocean water is a measure of how much water is in the global ocean.

As the waters of the ocean evaporate during an ice age, the water molecules (H_2O) that contain the lighter ^{16}O isotope evaporate more readily than the water molecules that contain the ^{18}O isotope. The water left behind in the oceans of the ice ages, therefore, has a greater abundance of ^{18}O compared to that found during interglacial periods. The shells of microorganisms found in the oceans, both plant and animal life, contain more of the ^{18}O isotope during ice ages than they do during interglacials because the oxygen in the calcium carbonate of these shells is extracted from the water in which they live. When these microorganisms die and sink to the bottom of the ocean, they leave in the bottom sediment a record of the relative abundance of the two different isotopes of oxygen in their shells. We shall talk about the chemistry of shell building in Chapter 6.

The results of a determination of the abundance of ^{18}O relative to ^{16}O along the length of a core taken from the Pacific Ocean are shown in Figure 5.6. When this ratio is converted to volume of water in the global ocean and the length along

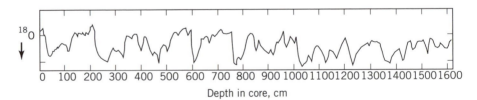

Figure 5.6 Variation in the amount of ^{18}O with depth in a sediment core from the floor of the Pacific Ocean. The full depth of the core (16 m) corresponds to a time of roughly one million years ago.

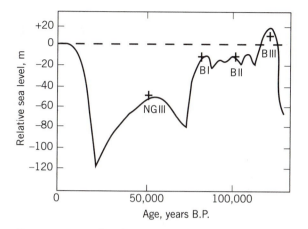

Figure 5.7 Sea level curve for the last 130,000 years as derived from the amount of ^{18}O relative to ^{16}O in roughly the first 2.5 m of the sediment core in Figure 5.6. This sea level curve is here compared with sea levels inferred from independent work in Barbados (B) and in New Guinea (NG).

the core is converted to age, we have the curve shown in Figure 5.7. Independent estimates of sea level have come from the tops of old coral reefs that once fringed the islands of Barbados and New Guinea. These estimates are also shown in Figure 5.7.

We thus have in Figure 5.7 a fairly detailed chronology of sea level and hence the amount of water locked up in ice and snow on land over the past ice age. One advantage of the ^{18}O record over the record of the sea level determinations used to calibrate it is that it is continuous and therefore captures the fluctuations such as those seen in Figure 5.7 that occur in an ice age. A second advantage of the ^{18}O record is that it goes back much further in time than the sea level determinations. This enables us to identify other glacial and interglacial periods and (although it is not very obvious from the record in Fig. 5.6) to establish that ice ages have occurred repeatedly at intervals roughly 100,000 years apart.

5.3 THE MILANKOVITCH THEORY OF THE ICE AGES

The cycles of glacial and interglacial periods over the past million years have spawned many theories to explain their occurrence. Milutin Milankovitch, a Serbian scientist, built on earlier work and proposed in the 1920s that these cycles were due to subtle variations in the solar radiation received by the earth that arise from changes in the orbit of the earth about the sun. Until the past decade or so, this theory was largely dismissed in the scientific community. The systematic documentation of the temporal characteristics of ice ages obtained from oxygen-isotope analysis of the sediments in deep sea cores have since rendered it a plausible explanation for the behavior of climate over these long time periods.

Milankovitch developed his theory regarding the occurrence of ice ages at a time before the record from deep sea cores for the timing of the ice ages was available. He postulated that diminished amounts of solar radiation received in high latitudes of the Northern Hemisphere in summer are the trigger activating the ice-albedo feedback loop (Fig. 5.5) that gradually brings on the ice age. These periods of reduced solar radiation in high latitudes he attributed to changes in the characteristics of the earth's orbit around the sun.

What is occurring in the ice-albedo feedback (Fig. 5.5) during the development of an ice age is a gradual increase in the area covered by ice and snow in high latitudes of the Northern Hemisphere. Milankovitch's theory specifically ascribes this to relatively cool summers in that part of the world associated with less than normal solar radiation received there. The idea is that, when the earth's orbital characteristics are such that they produce a long epoch of relatively cool summers, the snow that falls there in the winter doesn't melt completely. In this way, a snow-covered polar cap creeps ever farther southward, increasing the area covered and thereby maintaining the ice-albedo feedback loop.

Milankovitch recognized that three characteristics of earth motion vary. They are:

- tilt of the earth's axis of rotation
- eccentricity of the earth's orbit
- precession of the earth's orbit

It is difficult to visualize them. To simplify our presentation, we will look at each variation separately, as if the other two were held constant. Only at the end will we examine the cumulative effect of all three changes.

The *tilt* of the axis around which the earth rotates is what is responsible for the seasons. In Figure 5.8a the earth is shown oriented in such a way that the axis of rotation is perpendicular to the direction of travel of radiation coming from the distant sun. We used a similar figure in Chapter 3 (Figure 3.1) to make the point that the intensity of solar radiation incident on a sphere with this perpendicular orientation is highest at the equator and decreases toward either pole. That is because intensity is highest when the surface that receives the radiation is perpendicular to the direction from which the radiation is coming. The more inclined the receiving surface is, the less the intensity. Now in Figure 5.8b the earth is shown with the axis tilted 23.5° from the orientation that it has in Figure 5.8a. The season here is summer in the Northern Hemisphere. Clearly the tilt as shown has rendered any surface in the Northern Hemisphere less inclined (more nearly perpendicular to) the direction from which the radiation is coming. Thus the intensity of the radiation incident at any place in the Northern Hemisphere is greater in Figure 5.8b than it is in Figure 5.8a. The opposite is true in the Southern Hemisphere, where it is winter. Thus we have an explanation of why summer is warmer than winter.

If we were to increase the tilt further than it is in Figure 5.8b, then the radiation incident on the Northern Hemisphere would be even more intense. The amount of the tilt is 23.5° at present. It varies very slowly from 22° to 24.5° and back again over a time span of 41,000 years. The intensity in summer in high latitudes of the Northern Hemisphere is about 15 percent higher at the time of maximum tilt than it is at the time of minimum tilt. Cool summers in high latitudes

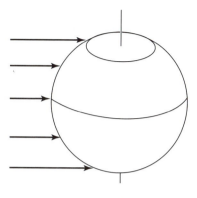

(a)

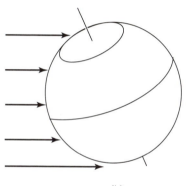

(b)

Figure 5.8 (a) Solar radiation incident on a spherical earth whose axis of rotation has no tilt; (b) same as (a) but with a tilt of 23.5°.

of the Northern Hemisphere are the condition for the development of an ice age. From what we have established here, the epoch most favorable for such summers is when the orientation of the earth's axis of rotation is near the minimum tilt.

The *eccentricity* of the earth's orbit varies over a period of about 100,000 years. At the present time the earth's orbit is nearly circular, with a slight eccentricity (that is, elliptical nature), as shown by one of the orbits in Figure 5.9. About 50,000 years ago, the earth's orbit was much more elliptical, and about 50,000 years from now it will be again. Such a highly elliptical orbit is also shown in Figure 5.9. We can visualize the earth as traveling around the sun on one or the other of these two orbits. The figure shows that there is a point on each orbit when the earth is closest to the sun. Astronomers call this "perihelion." A half year later, the earth will be at the point in the orbit farthest from the sun (so-called "aphelion"). At present, perihelion occurs in January and aphelion occurs in July. As a result, the intensity of solar radiation at the earth's distance from the sun, which in accordance with our calculation in Chapter 2 has the annual-average

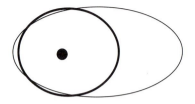

Figure 5.9 The present orbit of the earth about the sun is nearly circular (heavy line) compared to that 50,000 years ago (thin line).

value of 1368 W m^{-2} (the solar constant), at present is actually higher in January than it is in July. Because the present orbit is nearly circular, the difference is small, amounting to about 7 percent of the solar constant. It can be appreciated, however, that some 50,000 years ago, when the orbit was much more elliptical (the other curve in Fig. 5.9), this difference was considerably larger.

What we have just pointed out would indicate that some 50,000 years ago summers in the Northern Hemisphere would have seen solar radiation considerably less intense than it is at present. The truth of this statement is compromised by the added complication that the earth does not always reach the point in the orbit farthest from the sun (what we called aphelion) in July. In fact, the time of aphelion is at present very slowly regressing from the time of summer to the time of spring in the Northern Hemisphere. The cause of this regression is the final characteristic of the earth's orbit that needs to be considered: *precession*. This orbital characteristic is literally the wobble of the earth about its axis of rotation. You have seen precession whenever a top or gyroscope is spun. When a top is spinning fast, there is no precession, but as it slows down, the direction in which the axis of rotation points traces out a large circle with respect to the vertical.

Right now the earth's axis, as seen in Figure 5.10a, points toward the North Star. This axis of rotation will very slowly precess about a line perpendicular to the plane of the earth's orbit around the sun.

The proximity of the time of perihelion to the winter solstice is shown schematically in Figure 5.11a. The period of precession is 23,000 years. In 11,500 years, the earth's axis will point farthest away from the North Star (Fig. 5.10b). This in turn will mean that the time of perihelion will have advanced to July, as shown in Figure 5.11b, while aphelion will have regressed to January. What will prevail then is the opposite of what we said prevails today. Then the high latitudes of the Northern Hemisphere will receive more solar radiation during summer than will the high latitudes of the Southern Hemisphere during their summer.

We can summarize by stating that any one or a combination of the following favors the development of an ice age:

- the tilt of the earth's axis is less than that at present,
- the eccentricity of the earth's orbit is more than that at present,
- the time of aphelion is in Northern Hemisphere summer.

To visualize an outcome of combining these three factors, consider Figure 5.12. Here, three records of solar radiation are arbitrarily constructed with the same amplitude and with periodicities proportional to those observed for the tilt, eccentricity, and precession of the earth's orbit. Notice that the sum of these three records of solar radiation, each of which individually have well-defined periodic behavior, exhibits rather irregular behavior characterized by occasional periods of high solar radiation and other periods with low solar radiation.

Figure 5.12 is simply a mathematical exercise done to show that regular variations of tilt, eccentricity, and precession can combine to produce irregular variation. There is no correspondence between the irregular variation that we have created here and the particular irregular variation of sea level recorded in the core in Figure 5.6. One reason for this is that the three regular variations have different amplitudes, contrary to what we assumed in Figure 5.12. But allowance for this is not sufficient to explain lack of correspondence between the irregular pattern of the solar forcing and the irregular pattern of the response as recorded in sediment cores. Some current possibilities as to where the explanation may lie are the dynamics of the ice sheet itself and the geological process of the slow sinking and rebounding of the land under the ice sheet as it waxes and wanes.

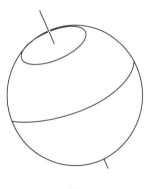

(a)

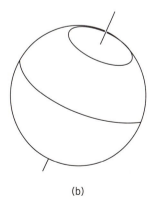

(b)

Figure 5.10 (a) At the present time, the axis of the earth points to the North Star. (b) The axis will point in the "opposite" direction 11,500 years hence.

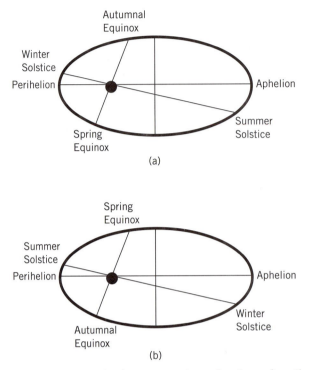

(a)

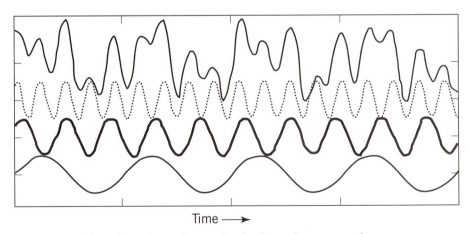

Figure 5.11 (a) At the present time, the time of perihelion occurs near the winter solstice. (b) In 11,500 years, the time of perihelion will occur near the summer solstice as a result of the precession of the earth's axis.

Time ⟶

Figure 5.12 Three hypothetical records of solar radiation are shown schematically as a function of time. The bottom solid curve depicts a record with a period of 100,000 years (eccentricity); the heavy solid line depicts a record with a period of 41,000 years (tilt); and the dashed curve indicates a record with a period of 23,000 years (precession). The top curve is the sum of these three records.

REVIEW QUESTIONS

1. How does the magnitude of the climate variations induced by volcanic activity and air-sea interactions compare to that induced by variations in the orbital characteristics of the earth?

2. What causes an ice age? How often do they occur? How long do they last? When did the last ice age reach it peak, and when did it end?

3. Why might the Eemian interglacial period that took place 120,000 years ago and the Holocene optimum period 5600 years ago be considered proxies for future periods with higher carbon dioxide concentrations than at present?

4. How and why do analyses of deep sea cores provide us with information on the temporal characteristics of ice ages?

5. Why does the depth of the ocean decrease during ice ages?

6. What is the ice-albedo feedback loop?

7. Why is the concentration of ^{18}O relative to that of ^{16}O greater in sediment cores during glacial periods?

8. Invent and describe a feedback process by which an ice age might come to an end.

INTERNET COMPANION

5.1 Paleoclimatology

- The NOAA Paleoclimatology Program provides access to many different types of paleoclimate records.

- The National Solar Observatory has made available pictures of the sun during periods of sunspot minima and maxima as well as long records of the number of sunspots observed.

5.2 The Ice Ages

- The change in the terrain of the earth's surface over the past 21,000 years is documented by the USGS Global Change Research Program. Short video clips are available to document the evolution of the ice sheets.

- The Illinois State Museum has an interesting interactive exhibit of the conditions likely to have been present in the Midwestern United States 16,000 years ago. Other information includes a short video clip of the retreat of the ice sheet over North America.

5.3 The Milankovitch Theory

- The Illinois State Museum also provides additional information, including other schematics, of the Milankovitch hypothesis.

CHAPTER 6

The Carbon Cycle

We learned in Chapter 4 that the amount of carbon dioxide now in the atmosphere is but a tiny fraction of the total that has come out of volcanoes over the ages. We remarked there that the process of removal of carbon dioxide from the atmosphere over time has been dissolution in ocean water and subsequent deposition of sediments containing carbon on the ocean floors. To this fact we now add that all carbon in living things on land and in the ocean has also come from the carbon dioxide in the atmosphere. In this chapter we will trace the paths of carbon as it passes from the atmosphere to these other *reservoirs*. We will see that the reverse paths are also open and that on these paths carbon passes from the reservoirs back into the atmosphere in the form of carbon dioxide as part of a global *carbon cycle*. One objective of this chapter is to make clear that the human activities of land clearing and fossil-fuel burning constitute a major disturbance of the natural carbon cycle.

6.1 THE PREINDUSTRIAL CARBON CYCLE

To trace the carbon cycle we must keep track of carbon atoms rather than carbon dioxide molecules. We do this by means of mass. The standard unit is 10^{12} kg of carbon atoms. Our unit of measurement of carbon dioxide concentration in the atmosphere has been parts per million (ppm) (see Fig. 1.1). It turns out that 2.1 $\times$ 10^{12} kg of carbon atoms are required to yield a concentration of 1 ppm carbon

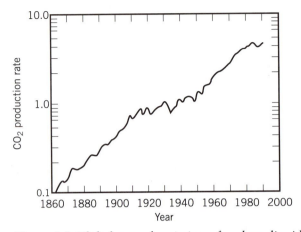

Figure 6.1 Global annual emission of carbon dioxide from the burning of fossil fuels expressed in 10^{12} kg of carbon per year.

dioxide throughout the global atmosphere. Thus the statement that the present concentration of carbon dioxide in the atmosphere is 360 ppm is the same as saying that the atmosphere presently contains 756×10^{12} kg of carbon in the form of carbon dioxide. The lower panel of Figure 5.3 shows that the "natural" level of carbon dioxide concentration before the era of extensive land clearing and industrial development that began around 1800 was 285 ppm. This concentration is equivalent to 600×10^{12} kg of carbon in the form of carbon dioxide. Since we are adopting 10^{12} kg as the unit, we will henceforth omit this factor and, for example, refer to the number just mentioned as 600 units of carbon.

The current rate at which the carbon dioxide concentration in the atmosphere is increasing is slightly less than 1.5 ppm annually, or 3 units of carbon. The result of a detailed inventory of global fossil fuel consumption is shown in Figure 6.1. The present rate of emission of carbon dioxide into the atmosphere from this source is 5 units per year. As we will see later, it is estimated that deforestation (land clearing) is the source of a further 2 units per year. With these numbers in front of us, we can see that less than one half of the carbon emitted into the atmosphere through human activities is actually accumulating there. In this chapter we will address the question of what constitutes the "sink" that is absorbing annually the remaining 4 units of the 7 units emitted.

The Land Portion of the Carbon Cycle

How carbon is exchanged between the atmosphere and the land portion of the biosphere is illustrated in Figure 6.2. The cycle begins with the intake of carbon dioxide by plants. By means of photosynthesis, this carbon dioxide plus water are used to create organic carbon molecules. The chemical reaction is given in Appendix A.4.

Some of these organic carbon molecules are consumed by the plant itself for energy. The chemical reaction in Appendix A.4 then runs in reverse, energy is

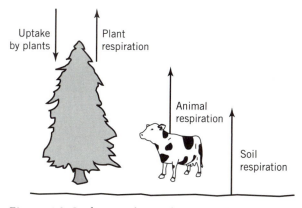

Figure 6.2 Carbon exchange between the atmosphere and the land portion of the biosphere.

liberated, and carbon dioxide is returned to the atmosphere. This return channel is labeled "plant respiration" in Figure 6.2. The difference between the amount of carbon taken in from the atmosphere and the amount returned by plant respiration is known as *net primary production* of organic carbon. The fate of this carbon is either to pass into the soil as litter or to be consumed by animals. As seen in Figure 6.2, "animal respiration" constitutes a further return of carbon dioxide to the atmosphere. The fate of the organic molecules that constitute the animals is to join the litter that goes into the soil. All this undergoes decomposition, which is for the most part respiration by bacteria and fungi, and this restores carbon dioxide to the atmosphere through the channel labeled "soil respiration" in Figure 6.2.

Carbon passes through this cycle at quite a rapid rate as shown by the numbers in Figure 6.3. This is the natural (that is, preindustrial) carbon cycle. The figure appears very complicated at first, largely because it also contains the marine portion of the carbon cycle, which we will examine in the next section. For now we focus only on the extreme lefthand side of the figure. There we see that plants on land globally assimilate atmospheric carbon dioxide at the rate of 100 units of carbon annually. Plant respiration returns 50 units annually, with the result that annual net primary production is 50 units. The arrow pointing vertically downward into the soil reservoir represents both litter fall and animal consumption. All this is respired, as represented by the arrow labeled "50 units" that is directed vertically from the soil to the atmosphere. What we have just said frames an important conclusion: The land portion of the natural carbon cycle is closed, that is, it constitutes neither a source nor a sink of carbon.

The reservoir contents are also given in Figure 6.3 in boldface type. It can be seen there that life on land (which consists mostly of plants) contains about as much carbon as the atmosphere, and the soil contains a little more than twice as much carbon as the atmosphere. Given that the atmosphere in these preindustrial times contains 600 units of carbon and that 100 units are exchanged annually with life on land, we see that the residence time of carbon dioxide in the atmosphere is 6 years for this process. (It is 3.4 years when the exchange rate of 74

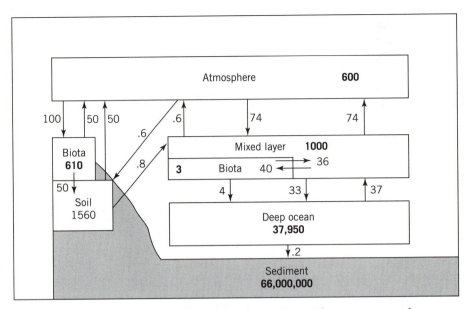

Figure 6.3 The carbon cycle before the industrial era. The amounts in the reservoirs are in units of 10^{12} kg, and the transfers between reservoirs are in units of 10^{12} kg per year.

units annually with the mixed layer of the ocean is considered.) This residence time is so brief that we can see that atmospheric carbon dioxide is a dynamic atmospheric gas that is intimately connected with life. The whole carbon dioxide component of the atmosphere literally passes through life on land every 6 years. On the other hand, this residence time of 6 years is very long compared to the residence time of 10 days that we deduced in Chapter 3 for atmospheric water vapor. This is the reason why atmospheric carbon dioxide is "well mixed" (concentration in ppm essentially the same everywhere in the atmosphere), while water vapor varies from almost zero to 4000 ppm depending on time and place.

The Marine Portion of the Carbon Cycle

The first stage of the marine portion of the natural carbon cycle is the exchange of 74 units of carbon annually between the atmosphere and the mixed layer of the ocean (Fig. 6.3). This global exchange occurs because carbon dioxide dissolves more readily in cold water than in warm water. Thus, for the most part, the exchange from the atmosphere to the ocean takes place in high latitudes and the reverse exchange from the ocean to the atmosphere takes place in low latitudes.

Unlike other constituents of the atmosphere, such as molecular nitrogen and molecular oxygen, carbon dioxide does not simply dissolve into ocean water. Instead, it dissolves and then undergoes the reaction given in Appendix A.4, which takes the carbon atom from the CO_2 molecule and puts it into a bicarbonate ion

(HCO_3^-). Because of this reaction, almost all the carbon dioxide dissolved in ocean water is in the form of bicarbonate ions. As chemical oceanographers do, we will refer to the carbon in bicarbonate ions as if it were all in the form of CO_2; that is, we simply call it dissolved carbon dioxide.

Just as atmospheric carbon dioxide is the source of all carbon for life on land, so also dissolved carbon dioxide is the source of all carbon for life in the ocean. The primary producers (photosynthesizing plants) in the ocean are the *phytoplankton*. The photosynthesizing reaction (the same as on land) takes place in the mixed layer, where there is light. The ocean herbivores are the oxygen-breathing *zooplankton*, which graze on the phytoplankton and, like animals on land, return carbon dioxide to the medium that they inhabit. Higher on the food chain are fish and other consumers. As shown in Figure 6.3, the reservoir of carbon contained in the biota of the ocean is rather small (3 units).

Figure 6.3 shows that the exchange of dissolved carbon dioxide with life in the mixed layer is not a closed cycle. Rather, there is a net export of 4 units of carbon annually from the biota of the mixed layer to the deep ocean. This constitutes what is known as the *biological pump*. Further inspection of Figure 6.3 reveals that the downward export of carbon by the biological pump is accompanied by a separate export of 33 units. This downward flow of 33 units and the upward flow of 37 units that balances it and the 4 units from the biological pump are achieved by the thermohaline circulation that was diagrammed in Figure 4.6. We will later examine this larger transport of carbon to and from the deep ocean. First we turn to an explanation of how the biological pump works.

When organisms in the mixed layer die, nearly all of their mass decomposes (is respired) in the mixed layer. The photosynthesis reaction then runs in reverse, and oxygen is consumed and carbon dioxide is released into the water. The remaining small fraction of organic carbon manages to sink out of the mixed layer and into the deep ocean. This constitutes for the mixed layer a net loss of carbon and a net gain of dissolved oxygen. Nearly all of this mass of dead organisms drifting down from the mixed layer subsequently decomposes in the deep ocean. In other words, nearly all the carbon raining down from the mixed layer into the deep ocean ends up as dissolved carbon dioxide there. However, a little bit of this mass manages to fall to the bottom and is buried in the sediments. This is the source of the organic carbon stored in sediments. Such burial of carbon constitutes for the whole ocean a net loss of carbon.

The burial of organic carbon also constitutes for the whole ocean a net gain of oxygen. To see this we return to the equation for photosynthesis in Appendix A.4. For every molecule of organic material (CH_2O) that is buried, an oxygen molecule remains behind in the ocean. Burial of organic material is thus accompanied by an increase in dissolved oxygen in the ocean. Since the amount of dissolved oxygen in the ocean is in equilibrium with (that is, proportional to) the amount of oxygen in the atmosphere, some of the increase in dissolved oxygen in the ocean gets shared with the atmosphere. This is the process, alluded to at the beginning of Chapter 4, that long ago built oxygen in the atmosphere up to its present level. That buildup has now ceased. This is because the present rate of increase of oxygen in the atmosphere is balanced by a loss due to the decomposition of organic material that was buried a long time ago on the ocean floor and has since been consolidated into rock and raised up as dry land by geological processes.

The transfer of organic carbon out of the mixed layer as just described is one component of the biological pump. There is another component. Both the phytoplankton and the zooplankton are prolific builders of tiny skeletal shells made of calcium carbonate ($CaCO_3$). The reaction by which the shells are constructed is given in Appendix A.4. The process of shell building takes place in the mixed layer, where these organisms live, and uses dissolved carbon dioxide.

Upon the death of the organism in the mixed layer, the shell survives to sink into the deep ocean. About 80 percent of the shells disintegrate there, which restores dissolved carbon dioxide to the ocean water. The remaining 20 percent make it all the way to the bottom and are buried there, and this constitutes for the whole ocean a net loss of carbon. This and the previously described burial of organic carbon are responsible for the trickle of 0.2 units of carbon per year into the sediment reservoir in Figure 6.3.

We now return to the thermohaline circulation of the ocean described in Chapter 4 in order to understand the other exchange of carbon between the mixed layer and the deep ocean that is represented in Figure 6.3 by the 33 units of downward transfer. As depicted in Figure 4.7, we can view the thermohaline circulation as starting from the sinking of cold, salty water into the deep ocean in the North Atlantic. This water moves very slowly as a river in the direction shown, mixing laterally with deep-ocean water at its sides, until it eventually comes to the surface in the Indian and Pacific oceans. As this deep water flows, there is a steady rain into it from above of dead organic matter and calcium carbonate, which is the output of the biological pump.

Combining the effects of the processes described in this section, we can see in Figure 6.3 that 33 units of carbon are carried down into the deep ocean and 37 units are upwelled. The sum of the contributions of carbon brought downward by the biological pump and sinking water is thus balanced by the carbon brought upward in the upwelling zones.

Figure 6.4 shows how dissolved carbon dioxide is distributed vertically in the ocean. The particular example shown here is for the North Pacific, but it can be considered for our purposes as representative of any location in the global ocean. The feature of this schematic that we wish to point out is that there is about 12 percent less dissolved carbon dioxide in the mixed layer (top 100 m) of the ocean compared to that in the deep ocean. This is a consequence of the presence of life in the mixed layer, as we now elaborate.

The biochemistry of living organisms requires atomic nitrogen (N) and phosphorous (P). These two elements are *inorganic nutrients*. They are very scarce in the ocean, as they are also on land. Phosphorous is a scarce element everywhere on earth and comes into the oceans by rivers, having been washed out of rocks by rain. The scarcity of atomic nitrogen (N) may seem strange in view of the fact that nitrogen is the dominant gas in the atmosphere and is abundant also as dissolved nitrogen in ocean water. But this nitrogen is molecular nitrogen (N_2), and it is difficult for molecular nitrogen to be converted into atomic nitrogen. The breakdown of N_2 requires certain kinds of bacteria. Some of these bacteria are found in the oceans while other bacteria on land create atomic nitrogen that even-

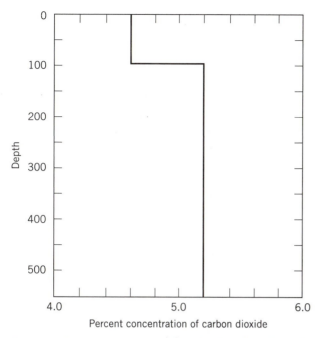

Figure 6.4 Concentration of dissolved carbon dioxide in percent of the total volume of ocean water as a function of depth (in meters).

tually washes into the oceans in the form of nitrate (NO_3^-). In any case, it is the lack of these inorganic nutrients in the ocean that limits the amount of life present there.

Analysis of dead organic material falling out of the mixed layer has revealed that it contains atoms of carbon, nitrogen, and phosphorous in the ratio of 100 to 15 to 1. The relative abundance of these three atomic species in deep water just as it begins to upwell into the mixed layer in the Pacific is in the ratio 800 to 15 to 1. What happens when this newly upwelled water arrives in the mixed layer is that life uses up both inorganic nutrients completely. Since oceanic life contains carbon, nitrogen, and phosphorous in the ratio 100 to 15 to 1, we can conclude that the leftovers consist of 700 of every 800 carbon atoms that come with the upwelling water. That is, there is 12 percent less dissolved carbon dioxide in the mixed layer than there is in the deep ocean.

This is the way the global ocean works. If the thermohaline circulation were to stop for some reason, then the rain of dead organisms would soon remove all the nitrogen and phosphorous from the mixed layer and put them in the deep ocean. Without the circulation to recycle these nutrients back up to the mixed layer, life in the ocean would cease.

The foregoing drastic scenario is somewhat oversimplified in that it does not take account of two regional aspects of the global ocean. One of these is the existence of places sufficiently shallow that the ocean bottom is actually in the mixed layer. Nutrients in dead matter that falls to the bottom in such regions get circulated back up into the water by the stirring processes that maintain the mixed layer. The second aspect is that of coastal upwelling driven by winds. This constitutes a recycling link from the deep ocean to the mixed layer that is independent of the thermohaline circulation.

It is literally true that inorganic nutrients are so scarce and life is so prevalent that nutrients are completely used up at those places where deep water in the thermohaline circulation upwells into the mixed layer. It remains to be clarified how there can then be life in the mixed layer elsewhere in global ocean. What happens is that life is short for the phytoplankton and zooplankton that constitute most of the ocean biomass. Recall that when organisms in the mixed layer die, only a small fraction of the material sinks into the deep ocean. The remainder decays in the mixed layer, and the nutrients are then available for use again. There is thus an intense recycling of nutrients within the mixed layer itself. This recycling is shown in Figure 6.3. Since 40 units of carbon enter the oceanic biota reservoir every year, we can say that the residence time of carbon in the oceanic biota category is about 0.1 year.

The pool of nutrients that came up in the upwelling zones drifts westward in the surface current that is returning water to the North Atlantic (see Fig. 4.7). There is a slow depletion of these nutrients in this return journey because of the continual drain of dead organisms into the deep ocean which constitutes the biological pump.

6.2 THE GEOCHEMICAL CARBON CYCLE

Let us collect together the content of all the carbon reservoirs shown in Figure 6.3.

Oceanic biota	3
Land biota	610
Atmosphere	600
Mixed layer	1,000
Soil and detritus	1,560
Deep ocean	37,950
Sediments	66,000,000

By far the largest reservoir of carbon is in the form of sediments. This reservoir includes not only ocean sediments but also rock formations on land whose chemical constituents were deposited as ocean sediments. All of the huge amount of carbon that is present in this reservoir came originally out of volcanoes, into the atmosphere, then into the ocean, and finally into the sediments. Carbon atoms making this journey for the first time constitute what geologists call "juvenile carbon." It turns out that an extremely small fraction of the carbon atoms found in the form of carbon dioxide in the atmosphere today is juvenile carbon. What this means is that the earth-atmosphere system has developed a way to recycle carbon. The complete cycle, involving as it does both geological processes and chemistry is referred to as the *geochemical cycle* of carbon.

The concept of reservoirs and residence time is central to understanding the ponderous geochemical cycle. The preindustrial atmospheric reservoir contains about 600 units of carbon (Fig. 6.3). Carbon dioxide dissolves into the surface of the ocean at a rate of about 74 units per year and is returned at the same rate. Considering only this gaseous exchange process, the residence time for carbon in the atmospheric reservoir is 8 years. The mixed layer of the ocean contains 1000 units of carbon. It exchanges carbon with the atmosphere at a rate of 74 units per year. Considering only this exchange process, the residence time for carbon in the mixed layer is 14 years. The residence time of carbon in the deep ocean is about 1000 years (as we mentioned in Chapter 4). The exchange here is between the deep ocean and the mixed layer. This residence time is very long compared to those characterizing the exchange between atmosphere and mixed layer (8 to 14 years, depending on which reservoir we focus on). Therefore, the deep ocean sees the atmosphere and mixed layer as one and the same reservoir. These two reservoirs can adjust to one another much faster than the deep ocean can adjust to them.

So far we have considered the exchange of carbon between the atmosphere and the mixed layer of the ocean and the exchange between this two-member combined reservoir and the deep-ocean reservoir. We go now to the much slower cycle in Figure 6.3, where 0.2 unit of carbon enters the ocean each year and where an equal amount passes to ocean-bottom sediments. Since the deep ocean contains 37,950 units of carbon, the residence time of this process is 190,000 years. This is very much longer than the 1000 years characterizing the exchange between the deep ocean and the combined reservoir of atmosphere and mixed layer. So, in relation to the process of loss of carbon to sediments, the atmosphere, mixed layer, and deep ocean all respond as one reservoir.

The 0.2 unit per year of carbon that goes into ocean-bottom sediments is replaced at this same rate by carbon entering the global ocean in rivers as indicated in Figure 6.3. The process that puts this carbon into rivers is the *chemical weathering* of rock formations by rain. What enables rain to do this is that it is naturally acid because it contains dissolved carbon dioxide. The geochemical carbon cycle is completed by invoking the geologic mechanism of plate tectonics that ultimately moves onto land the rock formations made from ocean-bottom sediments. There they undergo chemical weathering, which supplies the 0.2 unit of carbon per year that enters the ocean in rivers.

6.3 THE PRESENT CARBON CYCLE

The Current Numbers

We now return to the original question posed at the beginning of the chapter. What has happened to the carbon emitted into the atmosphere by the burning of fossil fuels and deforestation? Figure 6.5 shows a schematic similar to Figure 6.3 except that it represents the carbon cycle at the present time. Notice first near the left margin that two new sources of carbon for the atmosphere have been included: 5 units of carbon added each year by fossil-fuel burning and an additional 2 units added by deforestation. Notice also that the atmospheric reservoir has increased by 156 units to 756 units during the industrial era.

The schematic presented here (Fig. 6.5) accounts for all the carbon introduced as a result of fossil fuel burning and deforestation during the industrial era. This

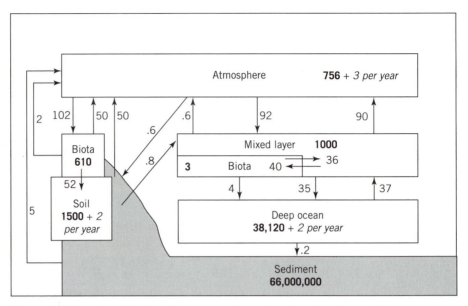

Figure 6.5 Present global carbon reservoirs and rates of exchange between them. The unit of measure for the reservoirs is 10^{12} kg while the rates of exchange are in 10^{12} kg per year.

has been estimated to be 320 units of carbon. According to Figure 6.6, this resides in the atmosphere (156 units) and in the deep ocean (170 units). What happens to the 7 units of carbon that are now being added each year as a result of the burning of fossil fuels and deforestation is that 3 units remain in the atmosphere; 2 units are fixed by plant life, possibly making their way into the soil; and the remaining 2 units are dissolved into the mixed layer and pass into the deep ocean via the thermohaline circulation.

This perturbation of the natural carbon cycle leads us to a new application of the concept of residence time. It can be considered loosely to be the time that a reservoir takes to come into equilibrium after it has been disturbed by changes in another portion of the system. Consider this year's 7 units of carbon dioxide input to the atmosphere. It will take roughly 11 years (1000 units divided by 92 units per year) for the mixed layer of the ocean to adjust to this new atmospheric

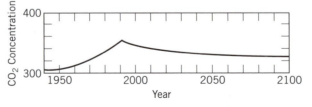

Figure 6.6 One estimate of the projected concentration of carbon dioxide in ppm if all emissions due to human activities were to cease now.

carbon dioxide. Otherwise stated, the mixed layer of the ocean has already responded to all of the carbon dioxide added to the atmosphere before roughly 11 years ago.

The deep ocean, with its much longer residence time of 1000 years, is just beginning to respond to all the carbon dioxide that has been added to the atmosphere since the Industrial Revolution began some 150 years ago. In effect, the deep ocean is responding to all the carbon dioxide that has been added to the combined reservoir of the atmosphere and mixed layer since that time, and is taking it in at a rate of 2 units of carbon out of every 7 units placed into the atmosphere. Finally, we see in Figure 6.5 that the land portion of the carbon cycle is absorbing 2 of the 7 units of carbon. How or when this stored carbon will make its way back into the atmosphere is a question upon which much current research is focused.

Deforestation

Our schematic in Figure 6.5 shows that 2 units of carbon per year are placed into the atmosphere as a consequence of deforestation. In a superficial way, deforestation is the most correctable aspect of human activity that would have a signficant effect on the concentration of carbon dioxide in the atmosphere. To correct the problem, it is simply necessary to grow trees at a rate faster than trees are being cleared. Or is it really so simple?

The tropical rain forests are the most productive in terms of the rate at which carbon is removed from the atmosphere and fixed through photosynthesis. The forests in middle latitudes are neither as large nor as fast in removing carbon. Grasslands, tundra, deserts, and concrete jungles are even less effective at removing carbon dioxide. As we have alluded to in the previous section, there is considerable uncertainty regarding how much of the carbon placed into the atmosphere through human activities is removed through storage in standing biomass or in the soils.

Most of the deforestation underway at present takes place in the tropical forests of South America, Southeast Asia, and Africa. The latest published figures from satellite surveillance indicates that 150,000 km^2 of tropical rain forest in the Amazon Basin fell to deforestation in the decade 1978–1988. This deforestation takes place as a result of a complex mix of economics, politics, and historical forces within the developing countries of the tropics.

There is another aspect of human activity that may be affecting the fixing of carbon in the biomass. As we mentioned in the previous section, life depends on the availablity of the inorganic nutrients nitrogen and phosphorous. We have altered the natural abundance of these nutrients with intensive applications of fertilizers. In the middle to late 1980s, roughly 70×10^9 kg of nitrogen were applied to fields, for the most part located in the middle latitudes of the Northern Hemisphere. The long-term effect of these additional nutrients is unknown.

Attempts have been made to estimate how much additional forest acreage needs to be planted in order to arrest the projected increase in the concentration of carbon dioxide. To take in 1 unit of carbon annually would require a new forest to be planted in temperate latitudes over nearly the area of the entire United States. In addition, as forests mature, they reach an equilibrium state between photosynthesis on the one hand and respiration and decay on the other and therefore cease to consume additional carbon dioxide. It is unlikely that planting trees will even temporarily reverse the effects of continued fossil fuel burning and deforestation.

How about reforesting the tropics? Estimates have been made that an area equivalent in size to the Amazon Basin could be replanted throughout the tropics (based on areas that were forested in the past or ones that are used for crops now). Most of the land to be reforested has been degraded by a reduction in amounts of phosphorous and other inorganic nutrients, so that the rate at which carbon could be fixed would be reduced compared to present natural tropical forests. Therefore, even the replanting of such a large area would not by itself remove more than an additional 1.5 units of carbon from the atmosphere per year. Based on both of these sets of calculations, we can conclude that the global warming problem will not be entirely solved by reforesting the earth.

The Projected Increase in Atmospheric Carbon Dioxide

Imagine for the moment that all the countries of the world were to unite to stop the emissions of carbon dioxide now. What would be the future concentration of carbon dioxide in the atmosphere? A projection is shown in Figure 6.6. It shows that the concentration would remain higher than that found in 1970 over the next hundred years. Why doesn't the carbon dioxide concentration return quickly to preindustrial levels? As we saw earlier, the carbon injected into the atmosphere over the past 150 years has upset the preindustrial equilibrium. The present situation is not one of equilibrium: The excess carbon introduced during the industrial era is being slowly absorbed in the deep ocean and in the land-biosphere reservoir.

The Intergovernmental Panel of Climate Change (IPCC) has developed four scenarios to span the range of possible alternative-energy-source measures for reducing carbon dioxide emissions. These can be summarized as follows:

- *Scenario A: Business as usual.* Few or no steps are taken to limit emission of greenhouse gases, and deforestation continues until tropical rain forests are depleted.
- *Scenario B: Natural gas.* Coal-intensive industries are converted to natural gas and deforestation is reversed.
- *Scenario C: Nuclear after 50 years.* Renewable energy sources (wind power, solar energy) and nuclear energy are dominant after about 50 years from now.
- *Scenario D: Nuclear now.* The shift to renewable energy sources and nuclear energy begins at the turn of the century.

Figure 6.7 summarizes IPCC estimates of how these four scenarios would affect the emissions of carbon dioxide into the atmosphere. Notice that none of these scenarios reduce the emission rate of carbon to zero. If nothing is done (Scenario

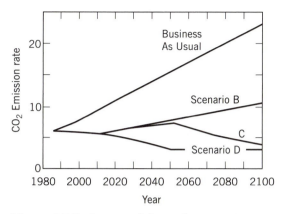

Figure 6.7 Estimates of the anthropogenic emissions (in units of 10^{12} kg per year) of carbon into the atmosphere projected to the year 2100 for the four scenarios described in the text.

A), carbon emissions will triple over the next hundred years while the modest (but economically painful) measures of Scenario B might put the carbon emission rate below 10 units of carbon per year.

The IPCC has also estimated what these four scenarios for carbon emissions would yield in terms of the concentration of carbon dioxide in the atmosphere (Fig. 6.8). If nothing is done (Scenario A), the projected concentration of atmospheric CO_2 will double from the current 360 ppm by about the year 2075. Only with the wrenching changes proposed in Scenario D would it be likely that the concentration of carbon dioxide would remain less than 450 ppm by this date. Recently, the IPCC has estimated that stabilization of atmospheric carbon dioxide at 450, 600, or 1000 ppm could be achieved only if carbon dioxide emissions drop to 1990 levels (7 units) by respectively, approximately 40, 140, or 240 years from now and drop substantially below 1990 levels subsequently.

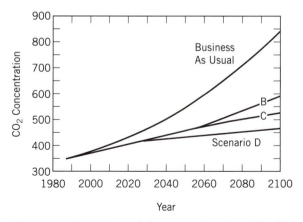

Figure 6.8 Estimates of the concentration of carbon dioxide (in ppm) projected over the next hundred years for the four scenarios described in the text.

REVIEW QUESTIONS

1. What are the processes by which carbon in the atmosphere cycles through the land portion of the biosphere? Through the marine portion of the biosphere?

2. How is carbon dioxide removed from the atmosphere?

3. What is the biological pump? How does it work, and what does it do?

4. How much carbon is presently being emitted into the atmosphere per year due to human activities? How much of this is presently accumulating in the atmosphere per year? Where is the difference between emission and accumulation going and by what processes does it get there?

5. What is the geochemical cycle of carbon in the earth-atmosphere system?

6. Why would you expect the Pacific Ocean to be richer in carbon than the Atlantic?

7. Why wouldn't the concentration of carbon dioxide in the atmosphere return to preindustrial levels if we stopped burning fossil fuels and deforestation for the next five years?

INTERNET COMPANION

6.1 The Preindustrial Carbon Cycle

- The Smithsonian Institution has an excellent interactive tour and exhibit called Ocean Planet. Information about the ocean portion of the carbon cycle is available.

6.2 The Geochemical Carbon Cycle

- The United Nations Environment Programme provides more than 90 fact sheets on global change issues, including the role of the oceans in the carbon cycle.

6.3 The Present Carbon Cycle

- The Intergovernmental Panel on Climate Change (IPCC) provides assessments on future emissions of carbon dioxide.

- The Rainforest Action Network is a forum for information on deforestation.

CHAPTER 7

※

Greenhouse Gases, Clouds, and the Radiation Balance

Water vapor is the predominant greenhouse gas in the atmosphere. Human activities have no direct effect on the amount that is present in the global atmosphere. The most important of the greenhouse gases whose amount in the atmosphere is affected by human activities are examined in the opening section of this chapter. We then introduce the idea of radiative forcing, which is a quantity that measures the effect that a specified change in any of these gases has on the radiation balance of the earth-atmosphere system. We include in this chapter an assessment in terms of radiative forcing of the possible effect on the radiation balance of small particles put into the atmosphere as a result of industrial activity. Finally, as we will see, the idea of radiative forcing can also be applied to measure the two competing effects of clouds on the global radiation balance. These two effects, introduced back in Chapter 2, are the greenhouse effect of clouds and the cloud-albedo effect.

7.1 THE OTHER GREENHOUSE GASES

Table 7.1 summarizes some of the characteristics of carbon dioxide and three other greenhouse gases that are affected by human activities. The reason that these other three gases are important, even though they are present in much lesser

Table 7.1 Characteristics of Several Greenhouse Gases.

	CO_2	CH_4	CFC 12	N_2O
Preindustrial concentration	280 ppm	0.8 ppm	0	0.288 ppm
Current concentration	360 ppm	1.7 ppm	500 ppt	0.310 ppm
Accumulation rate (annual)	0.5%	0.9%	4%	0.25%
Residence time	3 years	10 years	100 years	150 years

amounts than carbon dioxide, is that the molecules that compose them are very effective at absorbing infrared radiation. These gases are methane (CH_4), a representative chlorofluorocarbon (CF_2CL_2, also known by its informal abbreviation CFC 12), and nitrous oxide (N_2O). The concentration of methane has doubled since the Industrial Revolution began. Chlorofluorocarbons such as CFC 12 are not found naturally in the atmosphere, as evidenced by their absence prior to the industrial era. The last row of the table gives residence times in the atmosphere. The residence times of CO_2 and CH_4 are relatively short compared to that of chlorofluorocarbons. We will now review briefly some of the key aspects of each of these gases.

Methane

Methane is produced as a by-product of a number of anaerobic (oxygen-deficient) chemical reactions in the land portion of the biosphere. As summarized in Figure 7.1, the main sources of methane are emissions from natural wetlands, rice paddies, and animal digestive processes. Additional sources include methane emissions released as a result of drilling for natural gas and oil, burning of vegetation, the prolific eating habits of termites, and the anaerobic decay of organic material in landfills. Other minor sources have been lumped into a single category. Combined, all of these sources represent a flux of carbon into the atmosphere of 0.4 units of carbon per year.

The major sink for methane is oxidizing chemical reactions in the atmosphere. These reactions remove 0.345 unit of carbon in the form of methane each year, with another 0.022 unit removed by processes at the air-soil interface. This leaves us with an estimated net increase per year of about 0.033 unit of carbon in the atmosphere in the form of methane. This amount represents a tiny fraction of the 3 units of carbon gained each year in the atmosphere in the form of carbon dioxide (Fig. 6.5), but a methane molecule is much more effective in absorbing infrared radiation than a carbon dioxide molecule. Thus, methane is an important greenhouse gas, even though its amount in the atmosphere is small.

As seen in Figure 7.1, many of the sources of methane are independent of human activities. However, rice production and worldwide herds of cattle and sheep are significant sources of methane in the atmosphere. The upper panel of

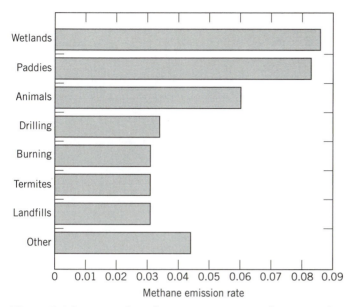

Figure 7.1 Sources of methane emissions into the atmosphere in units of 10^{12} kg per year of carbon.

Figure 7.2 shows concentrations of methane over the past 400 years measured from air in ice cores. The concentration of methane has increased very rapidly in the twentieth century. Methane concentration measured at stations scattered around the globe during the past decade are shown in the lower panel of Figure 7.2. The rate of increase of methane has slowed in recent years, apparently due to natural variations.

Chlorofluorocarbons (CFCs)

The CFCs are responsible for the depletion of ozone that is now occurring in the stratosphere. We will consider this in Chapter 9. Many governments signed the Montreal Protocol on Substances that Deplete the Ozone Layer in 1987 as a means of reducing the emissions of CFCs, and the principal producers of this substance agreed to a complete halt to production at the end of 1995.

CFCs were developed for industrial purposes such as refrigerants, spray-can propellants, and solvents. The chief advantage for these industrial applications is that CFCs do not react readily with other chemicals. The nonreactive characteristic of CFCs is their chief liability once they enter the atmosphere. They are destroyed by ultraviolet solar radiation in the stratosphere at a very slow rate, which gives them a long lifetime. The concentration of CFCs had increased at a rate in excess of any of the other greenhouse gases until the early 1990's. The Montreal Protocol has helped to slow the growth rate to near zero of CFCs, but other similar compounds continue to increase.

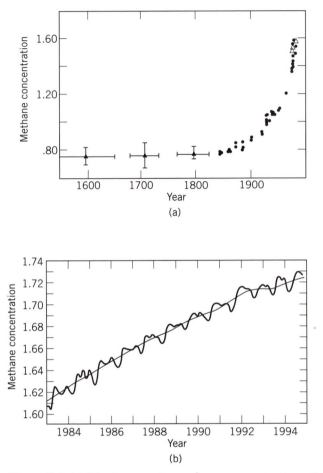

Figure 7.2 (a) The increase in methane concentrations (in ppm) based on air in ice cores. (b) Globally averaged methane concentrations (in ppm) during recent years.

Nitrous Oxide

Nitrous oxide is produced naturally from a variety of chemical reactions in land biota and in the oceans. Human activities such as combustion and the use of fertilizers have increased the amount of this substance in the atmosphere. Nitrous oxide is destroyed in the stratosphere by chemical reactions occurring naturally there.

7.2 RADIATIVE FORCING DUE TO INCREASE OF GREENHOUSE GASES

A large part of Chapter 2 was dedicated to the exposition of a simple model incorporating the process by which an increase of a greenhouse gas changes the radiative balance of the earth-atmosphere system. As we described there, an increase in a greenhouse gas raises the effective radiating level (ERL), with the result that the intensity of infrared radiation emitted to space is reduced. In this section we look at the question of how much that reduction is for a given increase in greenhouse gas concentration.

The amount of the reduction is termed the *radiative forcing* due to a specified increase in a particular greenhouse gas. When we speak of radiative forcing in this chapter, we will always say what the greenhouse gas is and what its increase is. In other words, the words "due to" form an essential part of the definition of radiative forcing. Radiative forcing must be calculated by means of numerical models whose details cannot be fully examined here. For the purpose of classification of the results, we point out here that such models are of two types. One is the three-dimensional climate model discussed in Chapter 3 that calculates many details of the flow pattern of the atmosphere and incorporates many physical processes other than the absorption of infrared radiation by the greenhouse gas that is being increased. What this type of model has to say concerning the response of the earth-atmosphere system to a given increase in the concentration of a greenhouse gas will be examined in Chapter 8.

The other type of model will be called a *radiative transfer model*. In this type of model the earth-atmosphere system is reduced to the simple model of Chapter 2, and the equations that govern the absorption of infrared radiation by the atmosphere are solved to yield the radiative forcing that results from a specified increase in greenhouse gas. Radiative forcing calculated by models of this type is the subject of this section.

Radiative transfer models have determined that the radiative forcing due to a doubling in the amount of carbon dioxide in the atmosphere is 4 W m^{-2}. If such a radiative forcing were to be imposed on the earth-atmosphere system now, the intensity of the outgoing infrared radiation in the global and annual average would be reduced from the present value of 237 W m^{-2} to 233 W m^{-2}. According to the simple model in Chapter 2, the earth-atmosphere would then warm up until the intensity of outgoing infrared radiation was restored to 237 W m^{-2}. We can easily calculate from the Stefan-Boltzmann Law that the amount of the warming would be 1°C.

Of course, the real earth-atmosphere system will not respond to the radiative forcing of 4 W m^{-2} in the same way as this simple model. We anticipate that the actual response to the radiative forcing will include a change in the atmosphere's water vapor content and in the type and amount of clouds that populate it. These things will have their own effect on the intensity of outgoing infrared radiation, with the result that the estimate of a 1°C global warming we have calculated may not be very representative of the amount of warming that will occur. Actual estimates of global warming will be treated in Chapter 8. For now we simply adopt 1°C as a guideline or very crude estimate of the warming that would correspond to the imposition on the earth-atmosphere system of a radiative forcing of 4 W m^{-2}.

Radiative forcing is the accepted measure of the relative importance to the global warming problem of the several greenhouse gases in Table 7.1. In the comparison of one greenhouse gas with another, the quantity of interest is the radiative forcing due to an increase in the greenhouse gas above some baseline level of our choosing. Consider Figure 7.3, which shows columns labeled by an interval measured in years. For example, in the interval 1765–1900, the year 1765 is the baseline year and the year 1900 marks the end of the interval over which we measure greenhouse gases that have been added to the atmosphere as a result of human activities. The column indicates that the radiative forcing due to the increase in CO_2 over that interval amounted to 0.4 W m^{-2}, and the radiative forcing due to the increase of CH_4 and all other greenhouse gases (except for the CFCs, which were not yet invented) amounted to a further 0.1 W m^{-2}. The radiative forcing due to increases in greenhouse gases in the interval 1900–1960 is given in the next column, and so on. Stacking all the columns in Figure 7.3 one above the other gives the radiative forcing due to the increase in greenhouse gases from preindustrial times to the year 1990. That number is 2.7 W m^{-2}.

We next consider the extension of this comparison of greenhouse gases into the future. Figure 6.8 at the end of the preceding chapter showed estimates of future atmospheric CO_2 emissions as the global economy grows at the expected rate. Estimates like this have also been made for other greenhouse gases. The calculated growth of radiative forcing (with preindustrial times as the baseline) for each of these gases under the business-as-usual (BAU) scenario, in which no specific measures are taken to limit emissions, is shown in Figure 7.4. As of 1990, the radiative forcing due to the sum of the gases was 2.7 W m^{-2}, as we noted.

Referring again to Figure 6.8, we can see that under the BAU scenario, CO_2 will have doubled from its 1990 value of about 350 ppm to 700 ppm by around

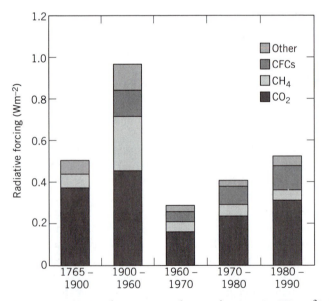

Figure 7.3 Contribution to radiative forcing (in W m^{-2}) due to increases in greenhouse gas concentrations since the onset of the industrial era. The sum of all of these contributions represents a change in radiative forcing since the onset of the industrial era equal to 2.7 W m^{-2}.

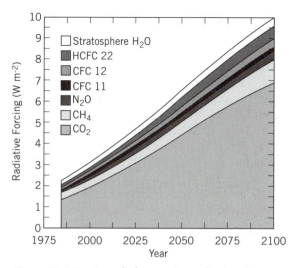

Figure 7.4 Projected changes in radiative forcing (in W m^{-2}) due to estimated increases in greenhouse gas concentrations using the business as usual (BAU) policy scenario.

the year 2075. We can confirm in Figure 7.4 that the radiative forcing due to CO_2 increase alone will have grown by 4 W m^{-2} over its present value by that year. If, however, we consider the effect of all the greenhouse gases, that growth of radiative forcing in the amount of 4 W m^{-2} will be realized sooner. According to Figure 7.4, that situation will arrive around the year 2035.

As we will see in the next chapter, where we consider estimates of global warming, the standard question that is asked of the models that estimate the warming is this: What is the global warming (that is, the increase in global- and annual-average surface temperature) due to a doubling of atmospheric CO_2? What this question is really referring to is the global warming due to an increase in all greenhouse gases (CO_2 included) that is *equivalent to* a doubling of CO_2 alone. We earlier saw that a doubling of CO_2 alone results in a radiative forcing of 4 W m^{-2}. The standard question can thus be rephrased more precisely as follows: What is the global warming due to a radiative forcing of 4 W m^{-2} arising from the projected increase in all greenhouse gases considered together?

7.3 AEROSOLS AND SULFUR

A reasonable question to ask is whether other industrial emissions could mitigate the effect of greenhouse gas emissions. We have already seen in Chapter 4 that the emission of sulfur dioxide during volcanic eruptions leads to the production of sulfate aerosol in the stratosphere. Industrial sources of sulfur dioxide also lead to the formation of sulfate aerosol in the troposphere. These tiny particles can in principle act to cool the earth-atmosphere system by reflecting visible solar radiation back to space and hence reducing the amount absorbed by the system. Before considering industrially produced sulfate aerosol in some detail, we review some other aerosols whose effect on the radiation balance of the earth-atmosphere system is less well known.

Aerosols such as sea salt that are hygroscopic (water-attracting) are referred to as cloud condensation nuclei (CCN). Airplanes and ships are significant sources of aerosols for the earth-atmosphere climate system. Contrails are clouds that form in the upper troposphere and lower stratosphere as a result of the CCN emitted by jet engines. Some contrails last only a few minutes, but under the right atmospheric conditions, contrails may be long lasting and cover a significant fraction of the sky. The formation of such "artificial clouds" could enhance the greenhouse effect by raising the local effective radiating level of infrared radiation emitted to space.

Ships emit large numbers of CCN into the lowest layers of the atmosphere. Under the right atmospheric conditions, the low-level clouds formed as a result of these additional CCN are noticeable in visible satellite images as long streaks that move with the prevailing winds. As discussed earlier, increased coverage of low clouds does not change the local effective radiating level of infrared radiation emitted to space. However, the increased cloud coverage does contribute to an enhanced cloud-albedo effect, increasing the amount of solar radiation reflected out to space.

While it is thought that the potential effects of changes in CCN concentration in the atmosphere are important, we cannot at the present time come up with an accurate estimate of their effect on the radiative balance of the earth-atmosphere system.

Sulfate Aerosol

Many chemicals found in the atmosphere have a cycle of generation, storage, and destruction that is as complex as the carbon cycle described in the previous chapter. One such chemical is sulfur. The atmospheric reservoir of sulfur is comprised for the most part of sulfate aerosol located in the lower troposphere. Some molecules emitted into the atmosphere that contain sulfur tend to react quickly with oxygen (that is, they oxidize rapidly) in the presence of sunlight to form the sulfate aerosol.

Figure 7.5 shows the sources of sulfur emitted into the atmosphere in units of 10^9 kg of sulfur per year. The total annual emission of sulfur is found by adding the subtotals in Figure 7.5 and is equal to roughly 150×10^9 kg per year. Since the atmospheric reservoir consists of only 1×10^9 kg of sulfur, the residence time of sulfur in the atmosphere must be only a couple of days. Sulfate aerosol in the lower troposphere tends to be swept out efficiently by precipitation. This is the source of *acid rain*, an anthropogenic problem of considerable significance in its own right.

The estimates of the natural sources of sulfur (the top-most three entries of Fig. 7.5) are subject to considerable uncertainty. Sulfate aerosol is also created in the stratosphere as a result of sulfur dioxide injected by volcanic eruptions. During the infrequent occurrence of a major volcanic eruption, the emission of sulfur can

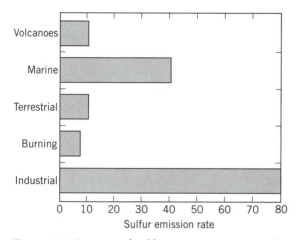

Figure 7.5 Sources of sulfur gas emission into the atmosphere (in 10^9 kg per year).

be much higher than that shown in Figure 7.5. Marine life is also a significant source of sulfur for the atmosphere. Phytoplankton produce a mixture of carbon and sulfur called dimethyl sulfide (DMS) gas. Roughly 90 percent of the sulfur emitted into the atmosphere from the ocean is in the form of DMS. Sulfate aerosol is created when the DMS reacts with oxygen in the atmosphere. Terrestrial plants and algae also emit DMS as well as other sulfur compounds such as H_2S.

The anthropogenic sources of sulfur can be sorted into two dominant categories: biomass burning and fossil-fuel combustion. The emission of sulfur due to biomass burning can be viewed as another impact of deforestation, and it constitutes a source of sulfur to the atmosphere that is comparable to that from volcanic emissions and terrestrial plant emissions.

The release of sulfur as a result of fossil-fuel combustion has increased considerably over the past 150 years. Emissions before 1850 were on the order of 3×10^9 kg of sulfur per year compared to the present value of 80×10^9 kg. In contrast to carbon dioxide, which has a residence time in the atmosphere of several years and is well mixed around the globe, the short-lived tropospheric sulfate aerosol is concentrated over and downwind of regions of major industrial activity. An estimate of the radiative forcing due to lower-tropospheric sulfate aerosol that originates from anthropogenic sources is shown in Figure 7.6. The largest radiative forcing is found in the industrialized areas of the world. Radiative forcing that represents a decrease in the intensity of infrared radiation emitted to space by the earth-atmosphere system (such as the 4 W m^{-2} due to a doubling of carbon dioxide) is, by definition, a positive quantity. Radiative forcing that represents a decrease in the intensity of solar radiation absorbed by the atmosphere, then, has to be a negative quantity, and that is what we see in Figure 7.6.

When the radiative forcing shown in Figure 7.6 is averaged over the globe, it contributes a radiative forcing of about -1 W m^{-2}. The data (Fig. 7.6) that go into this number are rather uncertain at the present time. According to Figure 7.4,

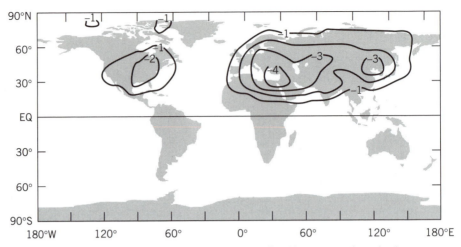

Figure 7.6 An estimate of the radiative effect of sulfate aerosol in the lower troposphere (in W m^{-2}). The negative values indicate that the aerosol acts to increase the solar radiation reflected back to space.

the radiative forcing provided by the increase in atmospheric carbon dioxide since preindustrial times stands at about $+ 1.5$ W m^{-2}. If the result shown in Figure 7.6 is accurate and all this radiative forcing is the result of anthropogenic sulfate aerosol, then much of the radiative forcing provided by increase of carbon dioxide since preindustrial times will have been canceled out by this global average radiative forcing of -1 W m^{-2}. That could explain why the global warming that is seen in the global- and annual-average surface temperature record (0.5°C) is so small.

If this turns out to be the explanation, then we would be faced with a policy dilemma. One strategy to minimize global warming would be to burn sulfur-enriched coal and thereby create more sulfate aerosol to reflect solar radiation to space and thus provide more negative radiative forcing. There are obviously some flaws in this strategy, one of which is that the acid-rain problem would undoubtedly become worse in the regions in which the sulfur is emitted. Stated in reverse, we can hypothesize that as attempts are made to curtail the acid-rain problem by reducing the burning of sulfur-rich coal, the global warming problem could be exacerbated. Furthermore, there is much uncertainty regarding how aerosols influence the amount and type of clouds that populate the global atmosphere. In the next section, we will consider the observational evidence that shows that clouds have an enormous influence on the radiation balance of the earth-atmosphere system.

7.4 RADIATIVE FORCING BY CLOUDS

The preceding sections were devoted to a description in terms of radiative forcing of the effect of increasing concentrations of greenhouse gases on the radiation balance of the earth-atmosphere system and the potential compensating effects of sulfate aerosol. The effect of clouds on the radiation balance can likewise be quantified in terms of radiative forcing.

Consider some local region of clear sky anywhere on earth. There the infrared radiation emitted to space is coming from the clear-sky ERL. If we now add to the region some mix of clouds whose tops are above the clear-sky ERL, then (as explained in Chapter 2) the radiation emitted to space from the region will be less than it was. Such a decrease constitutes a positive radiative forcing. Recall also from Chapter 2 that clouds have an additional effect: the cloud-albedo effect. Thus, when we begin with a clear sky and then add clouds, there will be an increase in the amount of visible solar radiation reflected back to space from the region. This corresponds to a negative radiative forcing, which acts to cancel and possibly even dominate the positive radiative forcing representing the decrease in emitted infrared. Given the competing effect of these two types of radiative forcing by clouds, we must carefully consider each of them.

We make these considerations in the context of the effect of the present *observed* global field of cloudiness on the radiation balance of the earth-atmosphere system. The data come from polar orbiting satellites that carried the instruments for the so-called Earth Radiation Budget Experiment (ERBE) in the 1980s. Each of the ERBE satellites carried instruments for measuring emitted infrared radiation and reflected visible radiation at points along a "scanning swath" under its orbit. Figure 7.7 illustrates the geometry of the situation. The narrow column with a single-pixel area at its base represents the field of view of a camera on board the satellite. The camera instantly records infrared radiation (x W m^{-2}, where x represents a number) emitted from some mix of clouds and clear sky in the region represented in the figure by a pattern of dots. This reading is compared with a reading obtained from the nearest region of clear sky along the same or, if need be, adjacent scanning swath. Let this reading be denoted by x_{cs}, where the subscript stands for "clear sky." We define the *infrared cloud radiative forcing* to be the quantity $x_{cs} - x$. Since x is less than or equal to x_{cs}, the quantity is positive or zero.

At the same time that the infrared observation takes place, another camera records the visible radiation (y W m^{-2}) in the region represented in the figure by the pattern of dots. This visible radiation consists of light reflected back from clouds together with light reflected from the earth and blue light scattered back toward space by the molecules of the atmosphere. Let y_{cs} denote the reading obtained by the same camera from the nearest region of clear sky. We define the *visible cloud radiative forcing* to be the quantity $y_{cs} - y$. Since y_{cs} is less than or equal to y, the quantity is negative or zero.

Let us abbreviate the term cloud radiation forcing as CRF. We have defined above the infrared CRF and the visible CRF. When different from zero, the former is always positive and the latter is always negative. The term *net* CRF refers to the sum of the two numbers. We can take the net CRF as measured by satellite at any location on the globe and average it for, say, an entire year. Such an annual-average figure at lots of points distributed all over the globe can then be averaged to yield the global average. The procedure is exactly the same as that which is

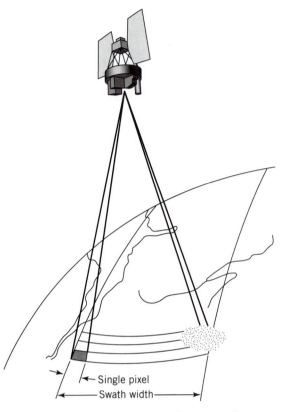

Figure 7.7 The cloud radiative forcing is determined by comparing the visible reflected and the infrared radiation emitted to space in cloud-free regions (dark square on the left edge of the scan) to that in adjacent cloudy regions (on the right edge of the scan).

followed to come up with the global and annual average surface temperature that we looked at in Chapter 1.

The result of applying this global averaging to net CRF is displayed in the following table:

Infrared CRF	31 W m^{-2}
Visible CRF	−44 W m^{-2}
Net CRF	−13 W m^{-2}

Here we easily see the cancellation effect: The net CRF is the difference between two numbers of comparable size. The negative sign of the net CRF indicates that the present effect of clouds is to make the earth-atmosphere system cooler than it would be if there were no clouds at all. What is particularly significant is the size of this number: It is a little more than three times larger than the (oppositely signed) radiative forcing that would be due to a doubling of atmospheric carbon dioxide (4 W m^{-2}). What's more, we can see from the table that the net CRF is the difference between two numbers that are fairly large compared to the differ-

ence itself. This signifies that the net is the result of a fairly delicate balance between the two different types of radiative forcing that clouds provide.

The existence of the balance just identified has implications for climate models
that are used to predict the response of the global earth-atmosphere system to a
doubling of atmospheric carbon dioxide. There is every reason to suspect that the
response of the system to the increased carbon dioxide will, among other things,
consist of a change in the characteristics of the global field of cloudiness. The
models have the task of calculating quite accurately both the infrared CRF and
the visible CRF in order to get the difference between these two comparably sized
numbers right. That the models are not yet up to this task will become evident
when we compare them against one another in the next chapter. The lack of
consensus as to which of a variety of ways currently used to simulate clouds in
climate models is best, together with the large net radiative forcing (-13 W m^{-2})
that clouds are presently observed to provide, jointly render simulation of clouds
the largest single uncertainty afflicting global warming predictions at this time.

We gain some insight into how the global population of clouds conspires to
produce the result in the foregoing table by breaking the global and annual average
down into some component parts. Figure 7.8 shows how net CRF is distributed
by latitude in the month of January and in the month of July. One thing that both
seasons have in common is that the net CRF in the tropics is very small. This is
made all the more remarkable by the fact that both the (positive) infrared CRF
and the (negative) visible CRF that go into the net are much larger numbers than
the global average figures that appear in the table. What is going on is that much
of the tropics is covered by clouds, and that provides a lot of visible CRF. The
widespread cloudiness in conjunction with the property that tropical clouds have

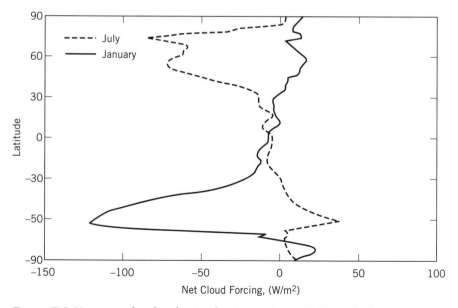

Figure 7.8 How net cloud radiative forcing varies with latitude during January
and July.

very high tops provides a lot of infrared CRF. As Figure 7.8 shows, there is almost complete cancellation of these two competing effects.

There are in Figure 7.8 two zones where the net CRF is large and negative. The more prominent of these is in the middle and high latitudes of the Southern Hemisphere. Large negative net CRF indicates considerable visible CRF and little infrared CRF. As we discussed in Chapter 2, this is the hallmark of a field of clouds with low tops. Such clouds do indeed predominate over the widespread cold ocean waters off the western coasts of Africa and South America. The atmosphere above the entire Antarctic circumpolar ocean is almost totally overcast with clouds of many types. Their tops are considerably lower than the clouds that populate the tropics, and this is what produces the very large negative net CRF in the zone between 50° and 70° south latitude. Six months later, there is negligible net CRF in the entire Southern Hemisphere. The zone of negative net CRF is now found in the higher latitudes of the Northern Hemisphere. There it largely reflects the presence of a lot of low- to middle-height cloud tops over the northern parts of the Atlantic and Pacific oceans.

R E V I E W Q U E S T I O N S

1. What are the three greenhouse gases other than carbon dioxide that are rapidly accumulating in the atmosphere as a result of human activities?

2. Why do CFC molecules last so long in the atmosphere while methane molecules do not?

3. What is the definition of the term "radiative forcing"? What (in $W\ m^{-2}$) is the radiative forcing due to a doubling of atmospheric carbon dioxide alone?

4. Why are one-dimensional radiative transfer models useful for studying feedbacks in the three-dimensional climate system?

5. Why do we say that doubling CO_2 could cause a reduction by $4\ W\ m^{-2}$ in the radiation emitted to space and, at the same time, we say that the atmosphere remains in radiative equilibrium?

6. If cloud top height changes in a given location, how does that affect: (a) cloud infrared radiative forcing, and (b) cloud visible radiative forcing?

7. If cloud amount changes in a given location, how does that affect: (a) cloud infrared radiative forcing, and (b) cloud visible radiative forcing?

8. What (in $W\ m^{-2}$) is presently the net cloud radiative forcing in the global and annual average?

9. Why would CCN aerosols emitted into the upper troposphere contribute potentially to an increased greenhouse effect while similar aerosols emitted into the lower troposphere would contribute potentially to a reduced greenhouse effect?

10. Why do sulfate aerosols last only a short time in the atmosphere? Why does the radiative effect of sulfate aerosols vary geographically?

INTERNET COMPANION

7.1   The Other Greenhouse Gases

- The NOAA/CMDL Nitrous Oxide and Halocompounds Division provides current information on the concentration of various chloroflourocarbons and nitrous oxide.

7.2   Radiative Forcing Due to Greenhouse Gases

- Information on the ERBE program is provided by NASA.

7.3   Aerosols and Sulfur

- NASA provides an additional fact sheet on aerosols and dimethyl sulfide.

- The NOAA/CMDL Aerosols group provides current information on aerosol concentration from reporting stations around the globe.

- A wide spectrum of information on acid rain is made available by the Environmental Protection Agency.

- Trends in acid rain are provided by the New Brunswick Precipitation Monitoring Program.

7.4   Radiative Forcing by Clouds

- Global maps of infrared and visible cloud radiative forcing demonstrate the complex role that clouds play in the climate system.

CHAPTER 8

Predicting Climate Change

We have reached the point where we can focus on a central question of this text. What will be the impact of rising greenhouse gases on global climate? The material in preceding chapters serves both as a foundation for understanding how such climate predictions are made and as a perspective for interpreting them. We start this chapter with a very brief review of what we have established. Then we examine those aspects of the earth-atmosphere system that are considered most important as sources of feedback in the process of global climate change. This feedback is contained within the models whose predictions of global change we then look at.

8.1 A Review

Here we review five basic observed features or characteristics of the earth-atmosphere system that have a bearing on global warming. We use the definition of global warming advanced in the opening chapter: an increase in the global-average surface temperature due to an increase in the amount of greenhouse gases in the atmosphere.

- Present amounts of greenhouse gases in the atmosphere, together with the present global distribution of clouds, are responsible for the fact that the global- and annual-average surface temperature is 34°C higher than it would be in their absence. This is the greenhouse effect.

- The global and annual average surface temperature is 0.5°C higher than it was a hundred years ago.
- The amounts of several greenhouse gases in the atmosphere are increasing as a result of emissions due to human activities. The greenhouse gas having the largest effect on the radiation balance of the earth-atmosphere system is carbon dioxide, and its amount in the atmosphere has increased by 30 percent since preindustrial times.
- When the amount of a greenhouse gas in the atmosphere increases, the average effective radiating level (ERL) moves to a higher altitude. Given that temperature decreases with altitude in the troposphere, this move results in a decrease of the intensity of infrared radiation emitted to space. Unless there are compensating changes in some other aspect of the earth-atmosphere system, an increase of temperature is required at this new altitude of the average ERL in order to restore radiative balance.
- Clouds are acting to cool the earth-atmosphere system at present, but this is the result of a delicate balance between the global cloud-albedo effect (cooling) and the global contribution that clouds make to the greenhouse effect (warming). This balance is certainly one aspect of the earth-atmosphere system that could change as increasing greenhouse gases act to raise the altitude of the average ERL. Such a change might drive the average ERL even higher (a positive feedback) or bring it back toward the level it had (a negative feedback).

8.2 THE FEEDBACK PROCESSES

Radiative transfer models that we referred to in Section 7.2 have also been useful for identifying a number of potential feedback processes that may enhance or diminish the effect of rising greenhouse gases on the radiative equilibrium of the earth-atmosphere system. Three-dimensional climate simulations are nearly as difficult to evaluate as the real climate system because the models incorporate so many physical processes. One-dimensional radiative transfer models can be used to evaluate specific physical processes selectively.

One of the most significant feedback processes results from the effect on temperature of the amount of water vapor in the atmosphere. We have said that human activity has little *direct* effect on the concentration of the largest greenhouse gas, water vapor. An *indirect* effect is shown in Figure 8.1. Rising carbon dioxide leads to an increase in the altitude of the ERL and reduced emission of infrared radiation to space. With other things assumed to remain the same, this requires that the temperature at the new position of the ERL increase in order to maintain radiative equilibrium. That same temperature increase will be realized at all levels in the troposphere right down to the surface, in order to maintain the property incorporated into the model of Chapter 2 that temperature decrease with altitude at the rate of 6.5°C km^{-1} both before and after the warming. However, the increase in temperature allows more water vapor to evaporate from the surface and to be held in the troposphere. A larger amount of water vapor in the tropo-

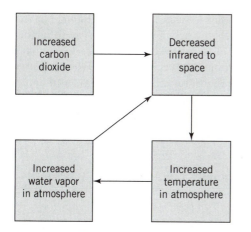

Figure 8.1 The water-vapor feedback loop.

sphere renders it more opaque to infrared radiation, which results in a higher altitude of the ERL and less infrared radiation emitted to space. Then the troposphere would have to warm even more to restore radiative equilibrium. In this manner, a *positive feedback* loop develops that can enhance the effect of increased concentration of carbon dioxide. Current estimates indicate a radiative forcing of between 2 and 3 W m^{-2} arising from water vapor feedback consequent to a doubling of carbon dioxide. Using the Stefan-Boltzmann Law, we can estimate that this would result in an additional warming of about 0.6°C.

We now introduce a parameter that specifies the effect of feedback processes such as the one we have just described. This is the *climate sensitivity parameter* that will be denoted by the Greek letter λ. It is the measure of how much the surface temperature changes for a given radiative forcing due to a specified increase in a greenhouse gas. Imagine for the moment that no feedback processes are acting in the earth-atmosphere system. It has been determined from one-dimensional radiative transfer models that in this case λ equals 0.33 °C per W m^{-2}. Because doubling the amount of carbon dioxide results in a radiative forcing of 4 W m^{-2}, this value of λ indicates that the surface temperature increase is 1.3°C. This is slightly larger than the rough estimate of a 1°C increase that we obtained in the preceding chapter when we used the Stefan-Boltzmann Law to deduce the temperature change resulting from a radiative forcing of 4 W m^{-2}.

Radiative transfer models indicate that when the water-vapor-feedback effect is included, then λ is equal to 0.45. The higher value indicates that the surface temperature is now more sensitive to (that is, responsive to) the radiative forcing that results from a doubling of carbon dioxide. Because the radiative forcing due to the doubling is 4 W m^{-2}, we conclude that the surface temperature increase is 1.8°C (4 × 0.45) as a result of the radiative effects of doubling carbon dioxide and allowing for water-vapor feedback.

A further positive feedback process has already been discussed in the context of the onset and demise of the ice ages. If rising carbon dioxide leads to warmer temperatures, then the present Antarctic polar ice cap may begin to melt. The ice-albedo-feedback process (Figure 5.5) may then work in reverse and lead to further warming and further melting. While several estimates of the ice-albedo feedback in the context of the global warming problem have been made with radiative transfer models, this process is so complex that it requires climate models that can simulate the atmosphere and oceans together. An assessment of cloud feedback (that is, radiative forcing due to a change in global cloudiness characteristics) also requires the use of three-dimensional climate models.

8.3   DOUBLING CARBON DIOXIDE IN CLIMATE SIMULATIONS

Validation of Climate Models

The best projections about what may happen to the climate system in the future are based on the results of simulations of the atmosphere with three-dimensional climate models. In order to have some confidence in what these climate models are predicting, we must first examine how well they simulate the present climate. Figure 8.2 compares the observed surface temperature variation with latitude to what is simulated by models developed by different research groups. This particular comparison is for the winter season of the Northern Hemisphere, and the temperature is therefore colder at the north pole than at the south pole. All the models can simulate with reasonable accuracy the variation of temperature with

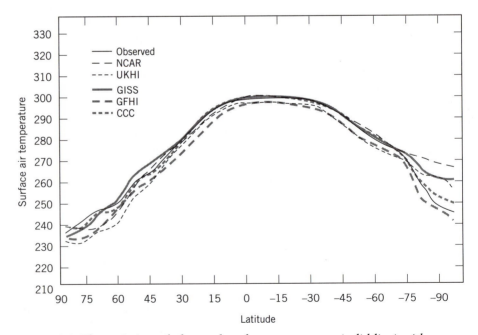

Figure 8.2 The variation of observed surface temperature (solid line) with latitude compared to temperatures obtained from several climate model simulations.

latitude, except over Antarctica (right edge of the figure). This failure is ascribed to the presence of the ice sheets in this region.

The ability of the same models to simulate the latitudinal distribution of precipitation is shown in Figure 8.3. The simulated precipitation exhibits larger deviations from the observed precipitation than those found for temperature. Some models predict much more rainfall near the equator compared to that observed; others predict less. This is a reflection of the fact that models used for weather forecasting or climate simulations have difficulty predicting clouds and, hence, precipitation.

Figure 8.4 compares the sensitivity parameter λ determined from simulations completed in the late 1980s with 14 different climate models. By definition, the larger the value of the sensitivity parameter, the more the global-averaged surface temperature changes in response to radiative forcing due to greenhouse-gas increase. We have arranged the values determined by each model from lowest sensitivity (on the left) to highest sensitivity (on the right). Thus, model 1 (with a value of 0.39 per W m^{-2}) is nearly three times less sensitive to radiative forcing than model 14 (with a value of 1.1°C per W m^{-2}). Assuming again that the radiative forcing due to a doubling of carbon dioxide is 4 W m^{-2}, the range of values indicate that the surface temperature could increase by as little as 1.6°C or by as much as 4.4°C.

The large range of climate-sensitivity parameter values indicates that there is considerable uncertainty as to its actual value. One way to determine a single value is to average all the available ones. This average value, as shown at the far

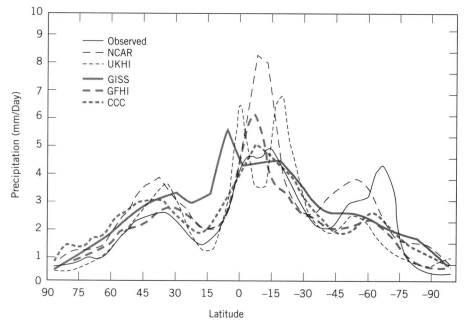

Figure 8.3 The variation of observed precipitation (solid line) with latitude compared to the precipitation obtained from several model simulations.

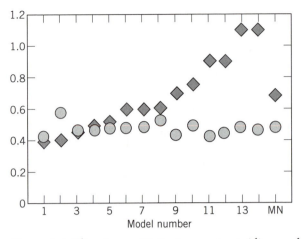

Figure 8.4 Climate sensitivity parameter λ (diamonds) and clear-sky sensitivity λ$_c$ (circles) for 14 different climate models. The respective averages (MN) of all of the values are shown at the far right. The values are in °C per W m^{-2}.

right in Figure 8.4, is 0.68°C per W m^{-2}. This value is higher than what was determined in the previous subsection, which indicates that cloudiness feedback renders the earth-atmosphere system more susceptible to a doubling of atmospheric carbon dioxide. With the value of 0.68 for the sensitivity, the change in surface temperature due to the doubling of carbon dioxide (radiative forcing of 4 W m^{-2}) is 2.7°C.

The sensitivity parameter for each model has also been evaluated for clear-sky conditions only. This procedure removes the contribution of cloudiness feedback. These values of clear-sky sensitivity (λ$_c$) are shown as circles in Figure 8.4. Notice that the range of 0.42–0.57°C per W m^{-2} is much smaller than the range of values of λ. All these models thus predict similar sensitivities when cloudiness feedback is removed. We can then infer that the wide range in sensitivity exhibited by the diamonds in Figure 8.4 is a consequence of the different methods for simulating clouds in different models.

There is one other disconcerting aspect of Figure 8.4. The first three models have values of λ that are less than their values of λ$_c$. This means that those models are more responsive to radiative forcing under a clear sky than under cloudy conditions. Put another way, these models suggest that clouds provide a negative feedback that would dampen the global-warming effect of rising greenhouse gases. On the other hand, the rest of the models have values of λ that exceed those of λ$_c$. These models suggest that clouds act to increase the sensitivity of the climate system to radiative forcing brought about by rising greenhouse gases.

The Simulations and a Consensus

Simulations with global three-dimensional climate models of the effect of increased concentration of atmospheric CO_2 follow a certain methodology, which we briefly review here. First, a model simulation referred to as the *control simulation* is made for present carbon dioxide content. In the early days, most models were run over a period of only a few months. The average of the weather simulated by the model over the last 30 to 90 days of the simulation was taken to represent the climate in the control simulation. It is also important to note that in these early models, only the atmosphere was allowed to evolve; conditions of the oceans and of the ice-covered regions remained fixed with time.

More recently, models of the climate system have incorporated the ocean. Recall that in Chapter 4, we divided the ocean into two parts: the mixed layer and the deep ocean. As we saw there, water from the mixed layer sinks into the deep ocean in certain places in high latitudes. It is now appropriate to remark that this process also removes heat from the atmosphere and stores it (for 1000 years) in the deep ocean. Until very recently the most advanced climate models included only the mixed layer of the ocean and not the underlying deep ocean. This situation was dictated jointly by lack of computing power and lack of knowledge of the deep-ocean circulation that an ocean model should simulate. We will consider in the next section what one simulation with a model that includes a deep ocean says about global warming. In this section we consider the results of a simulation where the only part of the ocean in the model is the mixed layer.

First the combined global atmospheric model and global mixed-layer ocean is run into the future with atmospheric carbon dioxide maintained at its present level. This is the control run. Then the combined model is run again with atmospheric carbon dioxide content at twice its present level. This simulation is called the *anomaly simulation*, since the imposed change in carbon dioxide content constitutes a departure from the present. The difference in surface temperature (or some other parameter) between the anomaly and control simulations averaged over some period represents the model's estimate of the global response to a doubling of carbon dioxide in the atmosphere.

Many doubling experiments have been performed with various different models, each consisting of a global, three-dimensional climate model coupled with a mixed-layer ocean. These experiments indicate that doubling the concentration of carbon dioxide would lead to changes in the global-average surface temperature between 1°C and 5°C. On the basis of these experiments, the general consensus among scientists working in this field is that a doubling of atmospheric carbon dioxide would lead to an increase in the globally averaged surface temperature at 2°C. This is lower than the figure of 2.7°C that corresponds to the average sensitivity determined by the models in Figure 8.4.

In this book we have adopted global-average surface temperature as the measure of global climate. However, the advantage of three-dimensional models is that they provide us with much more detailed information. In what regions of the earth will the temperatures rise the most? Who will get more rain? Who will get less? All these questions can, in principle, be answered by examining in detail the output from the simulations. However, just because it is possible to answer these questions in the context of such model simulations does not mean that the answers are right. One must keep in mind the considerable uncertainty that exists as a result of the models' inabilities to simulate cloud formation and other physical processes accurately. With this caveat in mind, let's now look at some of these answers.

What areas of the globe may experience the largest increases in surface temperature? Figure 8.5 shows one estimate of the regional changes in temperature resulting from doubling atmospheric carbon dioxide. This figure is for the Northern Hemisphere winter season. This particular model predicts an increase in global-average surface temperature of about 4°C, which is higher than the consensus figure of 2°C. The largest changes in surface temperature evident in Figure 8.5 occur in the polar latitudes, where in some places the temperatures are 8°C higher than in the control run. An increase in surface temperature of 8°C in polar latitudes would certainly have dire consequences for melting the ice sheets of Greenland and Antarctica and the sea ice of the polar seas.

Which regions will experience more/less rainfall? Figure 8.6 shows the change in precipitation predicted by the same climate model. The polar latitudes tend to have slightly higher precipitation in the anomaly experiment. Most of the central and eastern United States is predicted to have less rainfall than at present, while regions of the tropics (portions of Central America, India, etc.) are predicted to have greater rainfall. Unfortunately, each model predicts a different pattern of regional precipitation change resulting from doubling carbon dioxide. Thus, the simulations with many different models produce no basis for consensus on this particular aspect of climate change.

There is also no agreement among the model simulations as to other fields that are important for assessing the impact of rising carbon dioxide. For example, some models predict that the amount of soil moisture available during the growing season will decrease over portions of the central United States. However, none of

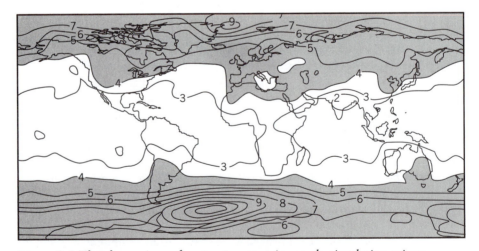

Figure 8.5 The change in surface temperature (anomaly simulation minus control simulation) predicted by a climate model as a result of doubling atmospheric carbon dioxide. Areas in which the change in temperature exceeds 4°C are shaded.

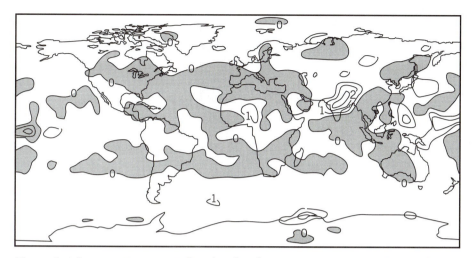

Figure 8.6 Same as Figure 8.5, but for the change in precipitation during the Northern Hemisphere summer season. Contours are shown at ±1, 2, and 5 mm per day. Areas in which precipitation decreases are shaded.

these models simulate accurately the processes that control soil moisture, and there are doubts about the significance of these results.

A Simulation That Includes the Deep Ocean

The simulations we have just described were undertaken with atmospheric models coupled to an ocean consisting only of a mixed layer. The addition of a deep ocean below the mixed layer is an advance that permits heat storage that has the potential to lessen somewhat the global warming of the atmosphere due to doubling of carbon dioxide and also the potential to alter the pattern of temperature change over the globe. In order for this fully coupled ocean-atmosphere model to evolve realistically, the imposed carbon dioxide increase must occur gradually rather than all at once, as was the case in the simulations just discussed. For this reason, the response of a model with a deep ocean is called the *transient response* to increased carbon dioxide. Many such simulations have been made. We will discuss here the one that uses the same atmospheric model from which Figure 8.5 was obtained. First, a model simulation was made in which carbon dioxide was held fixed at its present value (the control simulation). Then, another simulation was made in which the concentration in the atmosphere was increased by 1 percent each year (the anomaly simulation). About 70 years of simulated time must elapse in order for carbon dioxide to double, given this rate of annual increase.

Figure 8.7 shows the predicted change in global-average surface temperature as time progresses. This model simulation predicts that the global-average surface temperature would rise at a rate of about 0.35°C per decade to attain an increase of 3.5°C after 100 years. Notice that after 70 years, the temperature is expected

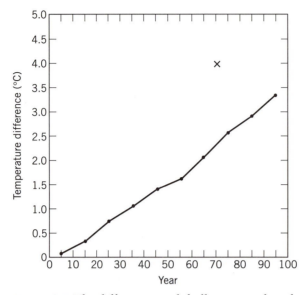

Figure 8.7 The difference in globally averaged surface temperature (in °C) between the transient response simulation, in which atmospheric CO_2 rises gradually, and the control simulation. The effect of an instantaneous doubling of CO_2 in a simulation without a deep ocean is shown for comparison by an x at 70 years, the time at which CO_2 concentrations at a rate of increase of 1 percent per year will have doubled.

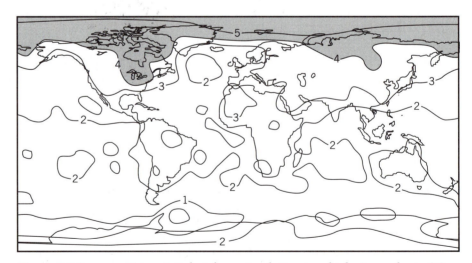

Figure 8.8 Same as Figure 8.5, but for a simulation in which atmospheric CO_2 increases gradually. The model in this simulation also incorporates heat exchange with the deep ocean. The contouring and shading are the same as in Figure 8.5.

to rise by only 2.3°C compared to the 4°C rise obtained in the simulation discussed earlier, in which the carbon dioxide is doubled instantaneously and there is no deep ocean. This result suggests that heat storage in the deep ocean could significantly moderate global warming.

Figure 8.8 shows the regional changes in temperature as a result of carbon dioxide doubling in this simulation. Compare the magnitude of the temperature changes here to those shown earlier in Figure 8.5. The change in surface temperature is now greater than 4°C only in the polar latitudes of the Northern Hemisphere. An even larger difference in the respective simulations is seen in high latitudes of the Southern Hemisphere, where the predicted 8°C warming (Fig. 8.5) has been reduced to 2°C. This relief is a consequence of storage of heat in the deep ocean, which is a particularly efficient process in the region of downwelling water that characterizes the model ocean in the region encircling the Antarctic continent. The model ocean also captures the phenomenon of downwelling in the North Atlantic that was highlighted in Figure 4.7, and this lowers the predicted increase of surface temperature there and in high northern latitudes in general.

The consensus view of the model simulations completed to date has been summarized by the Intergovernmental Panel of Climate Change (IPCC) as follows:

1. Confidence is higher in global-to-continental scale projections than in regional projections.
2. There is more confidence in temperature projections than in changes in the hydologic cycle.
3. Greater warming is expected over land than over the oceans, especially during winter.
4. Maximum warming is projected in the arctic during winter, with lesser warming during summer.
5. Precipitation and soil moisture may increase in high latitudes during winter.
6. The daily range in temperature is likely to be reduced.

All of these projected changes are associated with identifiable physical processes that can be simulated by these models.

8.4   RISK VERSUS UNCERTAINTY

Based on the information presented in the previous section, we must conclude that there is a definite risk that surface temperatures will be much higher by the middle of the next century than they are at present. Figure 8.9 shows estimates of how rapid the rise in global-average surface temperature could be as a function of the four policy scenarios of the Intergovernmental Panel of Climate Change (IPCC) introduced at the end of Chapter 6. If no significant changes are made to present policies (Scenario A), we could see an increase in the global-average surface temperature of 2°C within 50 years from now. The consensus figure of a 2°C increase due to a doubling of carbon dioxide is here seen to be reached before 2050, whereas Scenario A does not have carbon dioxide doubling until about 2075. The difference here is due to the fact that the 2°C increase is the consensus response

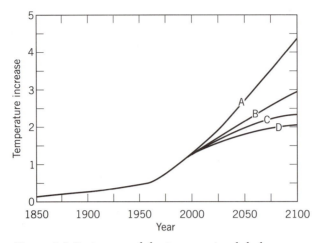

Figure 8.9 Estimates of the increase in global mean temperature relative to that in 1765 due to increases in greenhouse gas concentrations in the atmosphere. Scenarios A–D reflect different levels of controls on gas emissions.

to an increase of all greenhouse gases together acting as the equivalent of a doubling of carbon dioxide. The most recent estimate of the IPCC that takes into consideration projected future increases in aerosols suggest that this increase in temperature may be delayed until 2100.

There remain many uncertainties to resolve. Further work is required to estimate future emissions, the biogeochemical cycling, and radiative properties of greenhouse gases and aerosols. Clouds, oceans, sea ice, and vegetation remain incompletely represented in climate models. We turn now to weigh these uncertainties against the risks associated with greenhouse gases increasing as a result of human activities.

The Risks Associated with Global Warming

It has been said that humans have embarked on a climate experiment for which the outcome is uncertain. Assume for the moment that by the year 2010 the globally averaged temperature has increased by 0.5°C compared to 1990 values. There is now little question that global warming is underway. What might we expect? Beginning with our original definition of climate, we have focused throughout this book on average conditions, as represented by the global-average surface temperature. In this context, we have viewed weather as underlying climate, but we have never considered whether a change in climate might affect the nature of weather.

We have little hard information to go on, but it is certainly likely that many things could happen. Slight changes in the equator-to-pole temperature difference might lead to variations in the track along which the middle-latitude storm systems of the type seen in Figure 3.4 sweep across a region. This could cause regions to experience more frequent droughts or floods, depending on where the preferred storm track lies. The climate model simulations made to date exhibit too much variation from model to model to be able to determine changes in the preferred location of storm tracks with any degree of certainty.

What about hurricanes? A slight increase in the surface temperature of the oceans would provide conditions more favorable for the occurrence of hurricanes. A large number of tropical storms were observed in the Atlantic Ocean during 1995. Is this an indication of the type of climate likely to occur more frequently in the next century, or is it simply a result of natural climate variation? Climate simulations are incapable of predicting even the existence of hurricanes at this time, as a result of the poor horizontal resolution in the models.

Nearly every aspect of weather could be affected by global warming. Most of these changes would be so subtle that they would not be apparent. However, it is possible that global warming could affect the frequency of occurrence of the most violent weather (hurricanes, tornadoes, severe storms, etc.). In addition, subtle changes in the positions of the track of middle-latitude storms could lead to floods and droughts.

If global warming continues through the next century, what might happen to other aspects of the climate system? It is expected that rising greenhouse gas concentrations will lead to a warming of the ocean and slight thermal expansion of those waters. Coupled with the melting of glaciers and ice sheets, the best estimate of the IPCC is that the average sea level will increase by about .5 m ($1\frac{1}{2}$ ft) by 2100. Small islands and delta regions are most affected by the rising water levels. Higher storm surges could lead to significant erosion and flooding in many low-lying coastal areas. Paleoclimate data and model simulations suggest that the global thermohaline circulation could be weakened significantly by the melting of sea ice or ice sheets. A weakened thermohaline circulation would be less effective in alleviating warming by means of heat storage in the deep ocean.

What about impacts on terrestrial ecosystems? Climate change is expected to occur at a rapid rate relative to the speed at which many plant species, especially forest species, can respond. Most plants populate specific climate regimes that are controlled by latitude, elevation, and availability of water and nutrients. For mid-latitude regions, a global warming of 2°C has been suggested to be equivalent to a poleward shift in the local climate of 200–500 km and altitude shifts of 200–500 m. The composition of forests could change drastically and large amounts of carbon could be returned to the atmosphere during periods of high forest mortality while new species become established. Shifts in agricultural productivity are difficult to assess because regional changes in precipitation remain difficult to determine.

While people can adapt more quickly to a changing climate than natural ecosystems, people will be affected directly and indirectly by global warming. The vulnerability of coastal populations to flooding and land loss is especially high in many developing countries, such as Bangladesh. Extreme high temperatures (heat waves) may become more common and lead to higher mortality for people at risk to cardiorespiratory illnesses. Increased transmission of infectious diseases (malaria, dengue, etc.) as a result of broadened geographical and seasonal ranges of disease-bearing insects is likely. Whether we want to be or not, we are all part of this climate experiment. Our environment, economy, and lives are potentially at risk.

REVIEW QUESTIONS

1. Which of the four scenarios for future emissions do you think will be closest to what will eventually be observed? Why?

2. What is the predicted warming (in °C) due to a doubling of atmospheric carbon dioxide alone? By how much (in °C) is this warming increase due to water-vapor feedback? What is the consensus figure (in °C) as to what the warming will be due to a doubling of atmospheric carbon dioxide when all the feedback estimates are taken into consideration? What are these estimates?

3. What role will the ocean play in the global warming arising from a doubling of atmospheric carbon dioxide?

4. According to the "business as usual" scenario, how long will it take for atmospheric carbon dioxide to double from its present value?

5. What problems do climate models have in predicting the present climate?

6. Why do we need a control experiment as well as one (the anomaly experiment) in which CO_2 is allowed to increase?

7. Identify what factors you believe have the greatest risk associated with the global warming problem.

8. Identify what factors you believe have the greatest uncertainty associated with global warming.

INTERNET COMPANION

8.3   Doubling Carbon Dioxide in Climate Simulations

- The Model Evaluation Consortium for Climate Assessment provides interactive comparisons of many different CO_2 doubling model simulations.

- The Meteorological Research Institute of Japan provides video clips of a 70-year climate simulation in which CO_2 increases at a rate of 1 percent per year.

8.4   Risk Versus Uncertainty

- The National Greenhouse Advisory Committee of Australia has developed extensive background material on the global warming problem.

- The NOAA National Hurricane Center provides current and historical information on hurricanes that can be used to evaluate the variations in frequency and intensity of hurricanes with time.

- Access to several different viewpoints on the risks and uncertainties associated with global warming are available.

CHAPTER 9

Depletion of Stratospheric Ozone

According to the terminology that we introduced in Chapter 1, the *troposphere* is the layer of the atmosphere that extends from the earth's surface up to an altitude of about 10 km. Above this and extending to an altitude of about 50 km lies the *stratosphere*. Essentially all clouds and most greenhouse-gas components of the earth's atmosphere are contained in the troposphere. That is why in our examination of the greenhouse effect in Chapter 2 we were able to regard the top of the atmosphere as being the top of the troposphere. In this chapter we will look at the question of the effect of human activities on the ozone that resides in the stratosphere. This ozone has a relatively minor effect as a greenhouse gas. What it does is to protect the underlying earth from solar ultraviolet radiation that is harmful to life.

The environmental problem is that the ozone content of the stratosphere is decreasing, and this decrease has been shown conclusively to be the result of the industrial use of substances known as the chlorofluorocarbons (CFCs). These synthetic gases have accumulated in the troposphere and have found their way into the stratosphere. They have decomposed in the harsh ultraviolet light found there. Over three-quarters of the chlorine presently found in the stratosphere got there by this decomposition of the CFCs. Under conditions that we will examine in this chapter, ozone is very susceptible to chlorine. The remaining fraction (less than one-quarter) of present chlorine is natural and has long played a part in maintaining ozone in the stratosphere at its natural level. We mentioned that volcanoes represent a source of chlorine for the troposphere that is both large and intermittent, but this chlorine does not reach the stratosphere. It is in the form of hydrochloric acid, all of which gets washed out of the troposphere via the hydrologic cycle.

9.1   THE NATURAL LIFE CYCLE OF OZONE

We can quickly outline the natural life cycle of ozone before the advent of CFC-induced chlorine with reference to the simple schematic of Figure 9.1. Ozone is a gas that is produced in the upper stratosphere by the action of solar ultraviolet (UV) radiation on molecular oxygen. As was illustrated in Figure 2.3, UV radiation is characterized by wavelengths that are short (less than 0.3 μm) compared to visible light (roughly 0.5 μm). From this production region ozone moves downward and accumulates in the lower stratosphere, in what we will call the ozone storage zone. The storage there is temporary because ozone is continually leaking out of the bottom of this reservoir into the troposphere. There it is caught up in downward air currents associated with weather systems and brought into contact with the earth's surface, where it is destroyed. The troposphere thereby serves as a sink of stratospheric ozone.

The global distribution of ozone is illustrated by Figure 9.2. What we see here is the ozone storage zone. The troposphere below it (the ozone sink) appears as a clear region because it contains very little ozone. We see in this diagram another observed feature of the global ozone distribution: There is much more ozone over the polar regions that over the tropics. The diagram in Figure 9.3 can help us understand why.

Almost all ozone production in the upper stratosphere occurs in the tropics, because this is where the solar radiation, including its UV component, is most intense. As the newly produced ozone moves downward to the ozone storage zone in the lower stratosphere, it also moves outward toward the polar regions of both hemispheres. That is why more of it is stored there than in tropical latitudes. To find out what is driving this downward and poleward transport of ozone, we refer

		50 km
Ozone production	High stratosphere	
		25 km
Ozone storage	Low stratosphere	
		10 km
Ozone sink	Troposphere	
		Ground

Figure 9.1 Ozone is produced in the high stratosphere, stored in the low stratosphere, and destroyed in the troposphere.

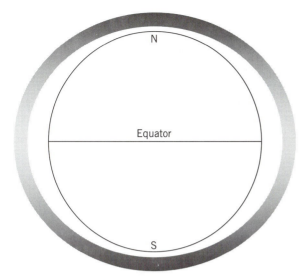

Figure 9.2 An illustration of the global distribution of ozone in the ozone storage layer; more ozone molecules are located near the poles than near the equator.

back to the discussion of the atmospheric heat engine in Chapter 3. There we said that the atmosphere tends to have rising motion in the tropics, poleward motion at high levels, and sinking motion in the polar regions. What we are seeing in Figure 9.3 is the uppermost portion of this global circulation pattern.

The Details of Ozone Production

In this discussion we can use the earlier simple diagram (Fig. 9.1) that illustrates in the global average perspective the zones of production, storage, and disap-

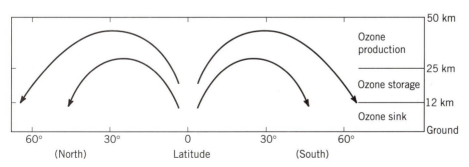

Figure 9.3 Most of the action in the ozone production layer takes place in tropical latitudes. From there ozone is transported poleward and downward by the atmospheric circulation that rises into the stratosphere from the tropical troposphere.

pearance of ozone. At the top of the production zone, UV radiation attacks molecular oxygen throughout the ozone production zone. We symbolically write this process of conversion of molecular oxygen (O_2) into atomic oxygen (O) as

$$\text{UV radiation} + O_2 \rightarrow O + O$$

The process is a statistical one in the sense that not all O_2 molecules that find themselves in this environment of ambient UV radiation are victims. The newly formed O atoms can combine with other O_2 molecules to form ozone molecules (O_3).

$$O + O_2 \rightarrow O_3$$

Now that ozone has been created, we look at two processes that destroy it in the very same production region. One process is that atomic oxygen, instrumental in producing it, can also eliminate it.

$$O + O_3 \rightarrow O_2 + O_2$$

In addition, ozone turns out to be susceptible to destruction by UV radiation.

$$\text{UV radiation} + O_3 \rightarrow O_2 + O$$

Again, the process is statistical: Not all O_3 that has been brought into this hazardous environment of UV radiation will succumb.

These four processes operate in the ozone production zone of Figure 9.2 and constitute the so-called "Chapman chemistry," named after the individual who advanced it as the explanation for the existence of the stratospheric ozone layer. We now know that there are some other chemical reactions that destroy ozone in the production zone. These do not concern us here, except for the one involving chlorine, which we will soon describe in detail. The net result of the four processes is:

1. to produce a mixture of O and O_3 in the region between 25 and 50 km, and
2. to absorb all incoming solar UV radiation in the region between 25 and 50 km.

As we remarked earlier in this section, there is a steady export downward of O_3 created in the production zone. The region into which it goes can be called the ozone storage zone, because there is so much of it there and because the UV that attacks ozone has been so depleted in the production zone that ozone is safe in the storage zone. Eventually, the ozone is exported to the troposphere, swept down to the ground, and destroyed.

The Classic Chlorine Catalytic Cycle

It is very risky to introduce chlorine into the ozone production region. If this happens, ozone can be depleted rapidly as follows

$$Cl + O_3 \rightarrow ClO + O_2$$

$$ClO + O \rightarrow Cl + O_2$$

We can describe this process as follows: In a mixture of ozone and atomic oxygen, chlorine will take one of each, form two oxygen molecules, and be free to repeat the chain reaction again. This is an example of a *catalytic reaction*. We will refer to this reaction as the *classic* chlorine catalytic cycle. It has been calculated that this reaction can occur 100,000 times before the chlorine atom involved bonds to other molecules and ends the process. Two reactions not given here eventually lead to the removal of the chlorine circulating in this catalytic cycle, placing the chlorine atom into the compound hydrochloric acid (HCl) or the compound chlorine nitrate ($ClONO_2$). These are called *reservoir species*. They are stable (non-reactive) compounds that, in effect, store chlorine.

Chlorine is being introduced into the stratosphere by means of the manufactured chemicals known as chloroflurocarbons (CFCs). As will be shown in section 9.3, the concentration of CFCs in the atmosphere has increased steadily over the past several decades. The dominant CFC is the compound CF_2CL_2, in which a carbon atom, two fluorine atoms, and two chlorine atoms are joined together. These CFC molecules decompose in the presence of the UV radiation found in the ozone production layer above 25 km. What occurs in the decomposition is the release of two chlorine atoms, each of which can enter the classic chlorine catalytic cycle.

This situation is not as bad as it may seem. It is true that chlorine released like this can and has decreased ozone in the ozone production zone. However, the vast majority of stratospheric ozone resides not in the production zone, where the decomposition of CFC molecules is taking place, but lower down in the storage zone. There is very little UV radiation to decompose CFCs down there, because most of the UV has been absorbed above, in the ozone production zone. Even if there were a source of chlorine down there, the classic catalytic cycle wouldn't take place because that cycle depends also on atomic oxygen being present, and that is found almost exclusively in the ozone production layer. Thanks to these mitigating factors, the worst case scenario has held that chlorine from CFCs would never amount to more than a 5 percent reduction in the total amount of stratospheric ozone. This was the prevailing thinking at the time of discovery (in 1985) of the ozone hole.

9.2   THE ANTARCTIC OZONE HOLE

Until the discovery of the ozone hole over Antarctica, it was widely believed that ozone in the ozone storage layer was immune from the effects of the chlorine introduced into the stratosphere by means of the global human production of CFCs. The ideas that formed the basis for this belief are easy to understand, and we review them again here. First of all, for reasons just noted, the classic chlorine catalytic cycle is effective in destroying ozone only in the ozone production region, where relatively little ozone resides. The fate of the chlorine that is liberated from

CFCs (after destroying a modest amount of ozone, which would amount to at most a decrease of 5 percent in the global total) is to become locked up in one of the two chlorine reservoir species ($ClONO_2$ and HCl). Just like ozone, these compounds are exported downward from the ozone production zone and into the ozone storage zone. There it was thought that chlorine, locked up as it is in reservoir species, could coexist with ozone.

Now we must make explicit mention of what process brings CFCs up from the troposphere into the upper stratosphere where the chlorine they contain can be liberated. This mechanism is none other than the circulation that is represented by the arrows in Figure 9.3. There is sufficient UV radiation at the altitude of the ozone production zone to break down the CFCs and release chlorine. This chlorine destroys ozone via the classic chlorine catalytic cycle. Eventually each chlorine atom will get locked up in one of the chlorine reservoir species. These species (HCl and $ClONO_2$) then get carried down into the ozone storage zone and outward toward the polar regions by the circulation.

The "ozone hole" or "hole in the ozone layer" is a region of very low ozone content that develops every year and lasts for a few months. The region where this occurs is centered on the South Pole and has an area comparable to that of the United States. The ozone hole is not evident in Figure 9.2, which shows only the natural distribution of ozone. The hole is thought to have first started forming in the 1970s, but was not noted until 1985.

A snapshot of the distribution of ozone in high latitudes of the Southern Hemisphere on October 1, 1995, is shown in Figure 9.4. The unit of measurement of ozone content is the "Dobson unit." We won't worry here about how much ozone corresponds to a Dobson unit's worth; what matters is the relative size of the numbers. The areas where the concentration of ozone is large are indicated by light shades of gray in Figure 9.4. This consists of most of the region between 30°S and 60°S. The ozone hole is evident poleward of 60°S, where the amount of ozone falls to less than 180 Dobson units.

Figure 9.5 shows the variation from month to month in the amount of ozone present over Antarctica, based on instruments on board satellites. According to the record shown in Figure 9.5, there is a distinct annual variation in the concentration of ozone over Antarctica. The minimum amount of ozone is measured either during October or November of each year while the maximum occurs a couple of months later in December or January. The variation in ozone in each of the first three years (1979–1981) is fairly similar from year to year, with minimum amounts around 300 Dobson units and maximum values around 400 units. A clear downward trend in the minimum amount of ozone is evident between 1982 and 1987. During this six-year period, the minimum ozone amount decreased from 290 to 190 Dobson units. The maximum amount of ozone observed near the beginning of each of these years is also significantly less than that observed at the same time during the first three years of the record. The annual variation in ozone during the 1990s appears quite regular and differs markedly from that measured a decade earlier.

Any explanation of the ozone hole must address the facts that (1) it is confined to a region of the globe centered on the South Pole, and (2) it develops rapidly in September, reaches a peak in October, and starts filling up again thereafter. A key

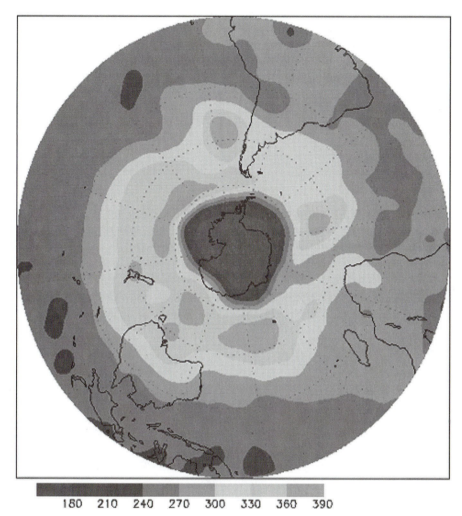

180   210   240   270   300   330   360   390

Figure 9.4 Distribution of ozone on October 1, 1995, in terms of Dobson units that indicate the total amount of ozone in the stratosphere in an atmospheric column of unit area. Lighter shades of gray correspond to larger amounts of ozone according to the scale at the bottom.

relation between this timing and geographical location is illustrated in Figure 9.6: The region is in constant darkness for several months before the ozone hole starts to form around the first of September.

We now outline the currently accepted theory for the formation of the ozone hole. First of all, it gets so cold during the Southern Hemisphere polar night that clouds form in the stratosphere. The most favorable place for these clouds is right in the middle of the ozone storage region, in the layer between about 15 and 25 km above the ground. In this layer we find not only ozone but also chlorine safely locked up (it was thought) in the reservoir species hydrochloric acid (HCl) and

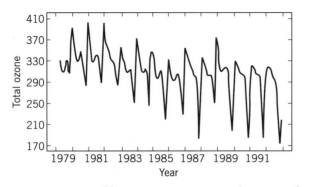

Figure 9.5 Monthly average ozone over the region from 70°S to the South Pole for the period from 1979 through 1992. The ozone is given in terms of Dobson units that indicate the total amount of ozone in the stratosphere in an atmospheric column of unit area.

chlorine nitrate ($ClONO_2$). Apparently, however, both species collect on the surfaces of the tiny crystals in the clouds. (These crystals are partly ice and partly frozen nitric acid.) On these surfaces the two reservoir species *react with each other*, and the net result of this is to liberate a chlorine molecule (Cl_2) into the atmosphere. All of this takes place during the months of the polar night.

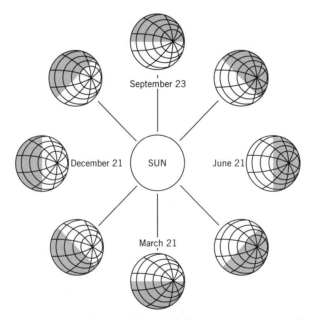

Figure 9.6 A view of the earth from a perspective above the South Pole. The region close to the South Pole receives no sunlight from March 21 to September 23.

By the time the long polar night ends, a large fraction of the chlorine atoms (Cl) that had been locked up in the two reservoir species is now back floating around in the ozone storage region over Antarctica in the form of molecular chlorine (Cl_2). In molecular form the two chlorine atoms keep each other busy, with the result that Cl_2 does not react with ozone. When the daylight comes back at the end of the polar night, solar radiation separates the chlorine molecule into two chlorine atoms.

The newly created chlorine atoms begin right away to destroy ozone. This time, the classic chlorine catalytic cycle is not the mechanism of destruction. That cycle requires the presence of atomic oxygen, and there is very little of that in the ozone storage region (as compared to the ozone production region where, as we saw, the classic cycle works very well). Rather, a different catalytic cycle comes into play. We do not need to go into the details of this new cycle. It is sufficient to note that, as in any catalytic cycle, the active ingredient can go around many times before it finally gets locked up in some other compound.

Eventually chlorine atoms get used up and the ozone destruction stops, usually by the end of October. Then the ozone hole evident in Figure 9.4 starts to fill in. Remember that the ozone hole is confined to an area roughly the size of the United States, but centered on the South Pole. The ozone that is filling up the hole is being brought from the equatorial regions (where ozone is produced) by the prevailing circulation in the stratosphere (Figure 9.3). Figure 9.5 suggests that this replacement of lost ozone is insufficient to bring the ozone maximum each year up to the level observed during the early 1980s.

An ozone hole of the severity found over Antarctica has not been observed to date in the polar region of the Northern Hemisphere. Due to some subtle differences in the atmospheric circulation in the respective hemispheres, the stratospheric polar night temperatures do not get as cold as they do in the Southern Hemisphere. This results in a less favorable environment for the formation of the stratospheric clouds in which the chemical unlocking of chlorine from the reservoir compounds takes place.

The threat of ozone-hole formation in the Northern Hemisphere doesn't end there, however. There is mounting evidence that sulfate aerosol particles do themselves serve as the host for chemical reactions that unlock chlorine from reservoir species. As described in Chapter 4, the eruption of the volcanoes El Chichón (1982) and Pinatubo (1991) resulted in an increase of stratospheric sulfate aerosol that spread over the globe. A modest, but observable, decrease in ozone followed. A further effect observed in the case of the Pinatubo injection was a dramatic liberation of chlorine in the polar region of the Northern Hemisphere in the spring of 1992. An ozone hole there was thought to have been averted only by the early transition of that entire region of the stratosphere to a summertime regime in that particular year.

This reveals yet another facet of the environmental hazard that CFCs pose. The injection of a large amount of sulfate aerosol distributed over the global stratosphere as a result of the eruption of a supervolcano in tropical latitudes is something that could occur anytime. We have loaded up the stratosphere with chlorine that is locked up in reservoir species, but a lot of it could be released by such an enhanced cloud of aerosol. Widespread ozone reduction could follow.

Some Consequences of Ozone Depletion

The Antarctic ozone hole represents a net loss of total atmospheric ozone over the year, and that is cause for concern. During the past decade the area covered by low ozone concentration associated with the hole has grown by a factor of more than four. Fortunately, the hole still remains confined to a relatively small fraction of the globe where there is essentially no human population. It is also fairly temporary, with the worst part lasting only a couple of months each year. But that minimum is drastic: a reduction in ozone of more than 50 percent. This reduction is ten times larger than the 5 percent reduction that was earlier thought to be the very worst that stratospheric chlorine could possibly do.

Why has there been such concern in the media regarding the ozone hole if it affects only a small part of the globe for only a few months? First, the development of the ozone hole in the past decade is a clear signal that human activity (in this case emission of chlorine compounds into the atmosphere) has created a demonstrable effect on the atmosphere. Second, there has been a (to date slight) downward trend in the ozone concentration over much of the rest of the globe during the past several years. Amounts during 1995 are detectably less than those only a few years earlier.

There is also increased concern over the link between the rising incidence of skin cancer in many countries in the Northern Hemisphere and exposure to UV radiation. The length of exposure to UV radiation has been established to be a significant risk factor for the development of skin cancer. The potential health risk has led the National Weather Service to develop and issue a daily UV exposure index for selected cities around the nation. This index can be found in newspapers and television broadcasts. It actually depends most significantly on factors that determine exposure to sunlight, such as time of year and cloud cover, rather than to the slight changes in UV radiation that result from present small variations in ozone concentration in the stratosphere outside of the Antarctic ozone hole.

9.3   OZONE DEPLETION AS A GLOBAL ENVIRONMENTAL ISSUE

As a global environmental problem, the matter of ozone depletion in the stratosphere has a different status from that of global warming: Steps have been taken to stop it. There has been impressive international governmental cooperation as well as cooperation between industry and government that will result in sufficient cutback in production of CFCs to allow chlorine in the stratosphere eventually to return to its natural level. We will review here the historical milestones of this environmental action on a global scale. In this country and several others, the production of CFCs was totally phased out by the end of 1995, and before the end of this century the number of CFC molecules in the global atmosphere is expected to begin to decline. As we will demonstrate here, however, the residence time of CFCs in the atmosphere is roughly 100 years. Thus, the decline will be slow, and there will continue to be an Antarctic ozone hole as well as a risk of serious ozone depletion elsewhere until well into the next century.

The Rise and Fall of CFC Applications

In the early part of this century, the first home refrigerators used ammonia as the fluid that removes heat from the interior of the refrigerator and deposits it in the air outside. A leak in the unit, which was a common occurrence in those days of early design, would result in the release of toxic ammonia gas into the household. With the interests of safety at heart one inventor in the 1920s came up with a "designer molecule" that was the first of the CFC family. It was not only nontoxic, it was also completely nonreactive. This latter property subsequently brought new applications. By the late 1960s, a CFC gas was serving as the propellant in spray cans all over the world. About the same time, a CFC gas became a mainstay in the rapidly growing industry of blowing plastics into foam. In this manufacturing process large amounts of CFC gas are emitted into the atmosphere. By the 1970s virtually every automobile rolling off the assembly line in this country was equipped with an air conditioner using CFCs in the refrigeration unit. A still more recent application has been the use of a CFC to clean computer chips.

Throughout the 1960s and 1970s, CFC gases continued to increase in the atmosphere. The first direct measurements of their presence in the troposphere were made in the early 1970s. Their continuing accumulation was generally acknowledged, but it was reasoned that their complete nonreactivity rendered them harmless there in all respects. That same nonreactivity assured that the CFC molecules would not remain in the troposphere where they were introduced, but would mix up into the stratosphere as well. This realization, together with the one that CFC molecules would release chlorine by decomposing in the ultraviolet light that they would be exposed to above the ozone storage layer, culminated in research that in 1974 convincingly demonstrated that CFC production constituted a threat to ozone in the stratosphere.

That the threat was really a rather qualified one came to light some years later when more detailed knowledge concerning the chemical reactions involved indicated that the potential global ozone depletion would probably not exceed 5 percent even with continued production of CFCs. This upper limit, as we noted earlier in the chapter, applies when the classic chlorine catalytic cycle is the only mechanism operating to destroy ozone. Discovery of the ozone hole and of the mechanism producing it lay in the future. There was nevertheless a considerable reaction to the fact that CFCs were an environmental hazard, yet were being sprayed into the atmosphere from cans. At the end of 1978, the use of CFCs as a spray-can propellant was banned in this country, and that ban has since spread to much of the world.

Meanwhile, CFC use in all other applications continued to grow. Instrumentation to monitor their accumulation routinely in the troposphere was deployed beginning in 1977. Figure 9.7 shows the concentration in the troposphere of the most abundant CFC, CFC 12. The overall trend during the past 20 years has been rapid increase in the concentration of CFC 12.

The concern that precipitated the banning of CFCs in spray cans in this country lived on. Around 1980 the United Nations Environmental Programme (UNEP) became the global sponsor for this concern. Throughout much of the decade, it

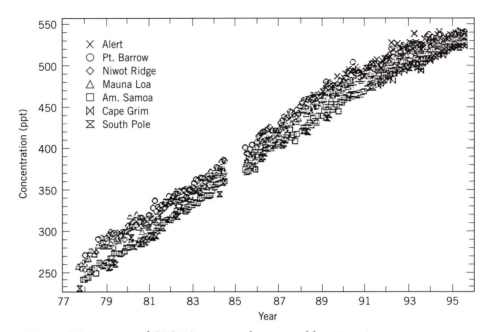

Figure 9.7 Amount of CFC 12 measured at several locations in parts per trillion.

presided over numerous decisions and policy-formulation meetings involving scientists and government representatives from the industrial nations. The culmination of this activity was the signing of the so-called Montreal Protocol in 1987. In this document some 24 nations agreed to set limits to production and use of CFCs. This beginning was modest, insofar as the terms were that the signatory nations would freeze CFC production at the 1986 level by 1989 and thereafter carry out a 50 percent reduction over the following decade. The agreement was tightened up in 1990 and again in 1992, so that all CFC production in the signatory nations would stop at the end of 1995.

What is the impact on our daily lives of this production phaseout? The most visible one is the change taking place in the system used to air condition cars. Roughly one-third of the CFC 12 produced in this country annually went for that application. A substitute has been developed by chemical manufacturers and is now being used as the working fluid in air conditioners installed in new cars. The substitute is "ozone friendly," meaning that it will not provide chlorine to the stratosphere. The majority of cars on the road today have air conditioning units that use Freon (a CFC brand name) as the working fluid. Those whose units have leaked know that a can of replacement Freon, once so cheap, is now expensive. Another impact is that recycling of CFCs has been widely legislated into effect. The worker who drains the working fluid from an air conditioning unit in a car, a home, or a supermarket in order to repair it has to save the fluid and put it back in the unit. A combination of very low price and the absence of any rule used to legitimatize the practice of simply letting the old fluid evaporate into the atmosphere as a gas.

9.4 THE FUTURE

The intergovernmental cooperation that formally began with the signing of the Montreal Protocol in September 1987 led to the complete phasing out of CFC production in many countries at the end of 1995. With that in place, we should see the CFC content of the atmosphere begin a slow decrease before the turn of the century. The reason that the decrease will be slow is that the residence time of CFCs in the atmosphere is long. The concept of residence time has been used in several applications in this book. Here we resort to it again to understand why this is so.

Just as we did for carbon dioxide, we can take the information in Figure 9.7, which is the concentration of the most abundant CFC in the atmosphere (in this case in parts per trillion), and convert it into the total number of molecules of that CFC in the global atmosphere. We arrive at the answer that there were roughly 5×10^{34} of these molecules in the atmosphere at the end of 1995. This is the reservoir. These molecules can be removed only by their disintegration in ultraviolet light in the upper stratosphere. As a parcel of air ascends into the upper stratosphere and moves poleward in the circulation depicted in Figure 9.3, the number of CFC molecules it contains steadily decreases as a result of their exposure to the ultraviolet light. There are accurate measurements of the concentration of this CFC and of the intensity of ultraviolet light at many points in the upper stratosphere. This information can be coupled with laboratory measurements of the rate of disintegration of this CFC in ultraviolet light of known intensity to yield the global destruction rate. It is now roughly 5×10^{32} molecules per year. This is the sink. Division of the reservoir content by this number yields a residence time of 100 years. The residence time of CFC 11, another CFC, is somewhat shorter, about 60 years.

Given these long residence times, the first 10 years or so of decline in the CFC content of the global atmosphere will certainly not be sufficient to end the Antarctic ozone hole. Worse yet, we will still be exposed over that decade or so to the threat of dramatic global ozone reduction consequent to a possible massive volcanic eruption in tropical latitudes. This reduction would last only a few years, until the stratosphere would cleanse itself of the volcanic aerosol on whose particle surfaces the chlorine-liberating reactions occur, but what would be the effect of that ozone reduction on the global ecosystem or on human health?

This provides a setting in which to reflect on the length of time that it has taken for a clear scientific basis supporting a regulatory measure on behalf of the global environment to be translated into international policy. We have seen that a period of some 7 years elapsed between the beginning of the laudable efforts of UNEP to get an agreement and the signing of the Montreal Protocol in 1987. Of course, for much of this period the consensus within the scientific community was that reduction of global stratospheric ozone due to CFCs would not exceed 5 percent. Then in 1985 came the announcement of the discovery of the ozone hole, which is characterized by a reduction much larger than 5 percent.

It took about two more years after this discovery to demonstrate that CFCs were in fact the cause of the ozone hole. We described in section 9.2 the currently accepted theory of how the CFCs do this. That theory was put forward in 1986. There remained then the task of getting the observational evidence that would confirm this theory. The final bit of this crucial evidence was brought to light just two weeks after the Montreal Protocol was signed in September 1987. It has been said that this meeting was ill timed. Had the full proof that CFCs were the cause of the ozone hole arrived somewhat before the Montreal meeting, the terms of the agreement that were put into effect in 1989 might have been stronger. In any case, 6 years later, at the end of 1995, we see the complete production phaseout. During that 6-year period, the CFC content of the atmosphere has gone up by some 25 percent. That, it is interesting to note, is nearly the same as the percentage by which atmospheric carbon dioxide has increased since the beginning of the Industrial Revolution.

Throughout this era of developing legislation to phase out the production of CFCs, it was known that they also constitute a greenhouse gas. No doubt this had an influence on the resolve to phase them out. However, the global warming threat by itself has not so far constituted grounds for the control of any anthropogenic greenhouse gas emissions.

There is no question that the action on the part of governments and industries that is embodied in the present agreement to move away from the use of CFCs is a direct result of clear articulation by the scientific community of firm scientific evidence. We hope that this book also makes apparent that there is at this time less scientific basis to underwrite control of carbon dioxide emitted into the atmosphere as a result of human activities. Next to water vapor, upon which there is no direct human influence, this is the dominant greenhouse gas. Its rise in the atmosphere, known to be due entirely to fossil-fuel burning and to land clearing, has been as well documented as the rise of CFCs. However, the effect of carbon dioxide and the other rising greenhouse gases on the warmth of the globe is just beginning to emerge in the observed temperature record.

The theoretical basis for the *expectation* of a global warming has been presented in this book in what we believe is its most elementary form. In its most advanced form, this theory is embodied in the computer models whose predictions we have examined. The consensus based on simulations with these models is that a doubling of carbon dioxide in the atmosphere will produce a warming of 2°C. Yet, as we have seen, the crucial process of the formation of all the types of clouds over many different regions of the globe is too poorly understood at present to be incorporated into these global computer models with any confidence.

We have in this chapter seen why the residence time of CFCs in the atmosphere is a long one. That global warming also has a long residence time can be inferred from material covered in earlier chapters. Once the atmosphere has experienced global warming in any amount, the mixed layer will within a few years experience a corresponding global warming. This leaves the deep ocean as the only remaining heat sink. Moving heat from the mixed layer to the deep ocean via the thermohaline circulation is a relatively slow process that could become even slower. That is because global warming can be expected to moderate the temperature of the cold, high-latitude surface waters that drive the thermohaline circulation. Thus, it has been said that by the time that we have firm evidence that global warming is taking place, it may be too late to take action to stop it.

REVIEW QUESTIONS

1. Explain why ozone is produced and destroyed at different levels in the atmosphere.

2. Why is ozone stored more at the poles and less at the equator?

3. Why does little short wavelength ($\lambda < 0.19$ μm) UV radiation reach the surface of the earth?

4. Describe briefly the classic chlorine catalytic cycle.

5. How do chlorine molecules enter the stratosphere?

6. Why does the ozone hole vary in intensity as a function of time of year?

INTERNET COMPANION

9.1   The Natural Life Cycle of Ozone

- Several extensive resources are available to explain the ozone cycle further, one of which is the NASA resource file on stratospheric ozone depletion.

9.2   The Antarctic Ozone Hole

- The NOAA/NCEP Climate Prediction Center (CPC) provides current and archived daily images of ozone for both the Northern and Southern Hemispheres.

- Further information on the networks used to measure ozone are available from the CPC Network for the Detection of Stratospheric Change.

9.3   Ozone Depletion as a Global Environmental Issue

- The NOAA/EPA UV Index for selected cities is available from many sources.

- Discussions of the scientific and economic implications of ozone depletion can be found in several locations.

Appendix

A.1 ABOUT NUMBERS

Some numbers are too large or too small to express conveniently by writing them out in full. Any number can more compactly be written in terms of powers of ten. This is called exponential notation. For example, a million (1,000,000) can be written as 1×10^6 and three billion (3,000,000,000) can be written as 3×10^9. Fractions can also be written in terms of powers of ten. A thousandth (.001) can be written as 1×10^{-3} (that is, $1/10^3$) while a billionth (.000000001) is 1×10^{-9}.

In this book, we often state numbers approximately. The term *roughly* means that the number lies within a range of 10 percent about that number. When we say that there are roughly 10 items, then there could be 9 or 11. The term *negligible* also has a specific meaning. It indicates that one number is several powers of ten larger than another. For example, carbon dioxide in the atmosphere constitutes only a tiny fraction of all of the gas in the atmosphere (roughly 3×10^{-4} of the total). So, we can say that carbon dioxide constitutes a negligible fraction of the atmosphere as a whole. However, carbon dioxide gas has a profound effect on the behavior of the earth-atmosphere system, even if its concentration is negligible compared to other gases.

A.2 ABOUT UNITS

The convention in all scientific disciplines is to use the SI (Système Internationale) units. Units are necessary in order to quantify distance, mass, time, and so on, as shown in the following table.

Quantity	Unit	Symbol
distance	meter	m
mass	kilogram	kg
time	second	s
temperature	kelvin	K
power	watt	W

Because of the large range of numbers that we will use, it is convenient to use prefixes such as kilo and milli. Kilometer (a thousand meters) and milligram (a thousandth of a gram) are common measures of distance and mass respectively. The following table lists prefixes that we will use to express the size of things:

Prefix	Size	Power of Ten
micro	millionth	10^{-6}
milli	thousandth	10^{-3}
kilo	thousand	10^3
mega	million	10^6

We will have an occasional need to convert quantities in SI units to the more familiar English system of units (miles, pounds, degrees, etc.).

Temperature in Kelvin can be obtained from temperature in degrees Celsius by adding 273. For reference, the melting temperature of ice is 273 K (0°C) and the boiling temperature of water (at sea level) is 373 K (100°C). To convert from degrees Celsius to degrees Fahrenheit, the following equation is used:

$$°F = 1.8 \times °C + 32.$$

Ice melts at 32°F and water boils at 212°F.

A.3 ABOUT GRAPHS

We show a number of figures in this book that convey how one quantity depends on another quantity. Consider Figure A.1, one of the central figures of this book. It shows how the globally averaged surface temperature has varied with time from 1860 through 1995. Now look at Figure A.2, which has exactly the same information, but the aspect ratio of the graph has been changed: The time axis (left-to-right direction) is compressed while the temperature axis (up-down direction) is expanded.

By means of changing the aspect ratio of this figure, the temperature rise has been made more dramatic. The temperature increase of roughly 0.5°C observed over the past 130 years isn't any different in Figure A.2 than it is in Figure A.1, but the impact is quite different. Figure A.2 lends itself to a more alarming view of climate change, even though the information content is the same.

A.4 SOME BASIC CHEMISTRY

We use some very basic concepts from chemistry in this book. As you are probably aware, an atom is composed of a nucleus around which electrons orbit as shown

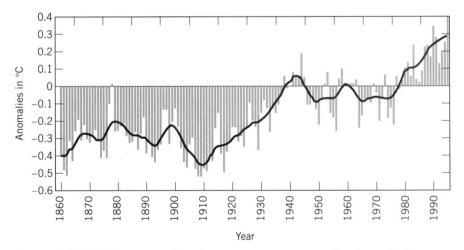

Figure A.1 Globally averaged surface temperature anomalies from 1860 to 1995 (in °C) relative to the 1961–1990 average.

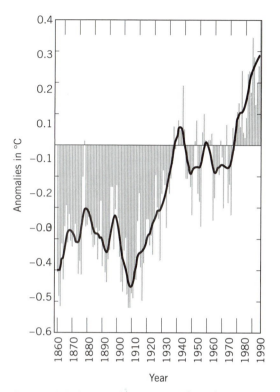

Figure A.2 *Same as Figure A.1 but the aspect ratio has been changed.*

schematically in Figure A.3. The mass (weight) of the electrons is negligible compared to the mass of the nucleus, which is composed of protons and neutrons. The mass of a proton is the same as that of a neutron. Consider now an oxygen atom. It has 8 protons and 8 neutrons in its nucleus with 8 electrons in orbit around the nucleus. Rather than drawing a diagram with a nucleus and 8 electrons circling around it, it is more convenient to use notation that summarizes the characteristic of the oxygen atom as ^{16}O. Here, the superscript indicates the number of neutrons and protons combined

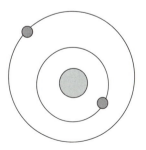

Figure A.3 *Two electrons orbit around a nucleus composed of protons and neutrons.*

in the nucleus. We could get away with writing it simply as O, except that there are three different forms of oxygen.

$$^{16}O \quad ^{17}O \quad ^{18}O$$

These different forms of oxygen are called oxygen *isotopes*. The first one is the most common form, while the other two are rarely present in amounts in excess of a few tenths of a percent. The differences between the three isotopes of oxygen are the number of neutrons found in the nucleus; the ^{18}O isotope has two additional neutrons, so that it is 12 percent heavier than the ^{16}O isotope.

A molecule is an aggregate of two or more atoms. The molecule carbon dioxide is made up of one carbon atom and two oxygen atoms. Either the molecule or the gas composed of the molecules can be denoted by the symbol CO_2. Note that with this notation, we don't say which isotope of oxygen is present or, for that matter, which isotope of carbon.

In a few places in this book we write chemical reactions that describe how different atoms and molecules react with one another. An example is the process of *photosynthesis* by which plants consume carbon dioxide (CO_2) and water (H_2O) and produce molecular oxygen (O_2) and a carbohydrate unit that is a fundamental part of glucose (sugar).

$$CO_2 + H_2O \rightarrow CH_2O + O_2$$

In the process of *respiration*, the reaction runs from right to left, putting the carbon atom in the carbohydrate unit back into a carbon dioxide molecule.

Most of the atmosphere consists of the gas nitrogen, whose molecule is denoted N_2. Some molecules of atmospheric nitrogen find their way into the ocean to constitute what is called dissolved nitrogen. Atmospheric carbon dioxide molecules also enter the ocean to constitute dissolved carbon dioxide. What distinguishes carbon dioxide from all other dissolved atmospheric gases in the ocean is that it is chemically active in water.

$$CO_2 + H_2O \rightarrow H^+ + HCO_3^-$$

The products of the reaction are a hydrogen ion (H^+, which is a hydrogen atom whose sole electron has been removed) and a bicarbonate ion (HCO_3^-, which has taken that electron). A small fraction of the bicarbonate ions undergo a further reaction, but this fact is of no importance to us here. In summary, what we have seen here is that the carbon in carbon dioxide dissolved in ocean water actually resides in bicarbonate ions.

In another relevant chemical reaction many marine plants and animals make shells.

$$Ca^{++} + 2HCO_3^- \rightarrow CaCO_3 + H_2O + CO_2$$

The calcium ion (Ca^{++}) is a calcium atom that has lost two of its (negatively charged) electrons and hence has a net positive charge of two units. The bicarbonate ion (HCO_3^-) consists of a hydrogen atom, a carbon atom, and three oxygen atoms, with one electron missing from the combination.

There is no net gain or loss of charge when, as in the above reaction, two bicarbonate ions combine with one calcium ion. The reaction yields calcium carbonate ($CaCO_3$), which is the material of the shell, plus water and carbon dioxide. After the organism that has made the shell dies, the shell may dissolve back into its original chemical constituents. In this case the reaction above proceeds from right to left instead of from left to right as we have shown it here.

A.5   Access to the Internet Companion

As discussed in the Preface, the purpose of the Internet Companion is to provide access to text and graphics that expand on the concepts presented in the text. A brief introduction to some of the terminology required to use the Internet Companion is presented here.

The Internet refers to the communication links that exist between computers connected together as part of national and international networks. Access to information created at one location can be obtained in several ways; one method used by many institutions is the World Wide Web. The Web requires servers (providers of information) and clients (seekers of information). The Internet Companion relies on a Web server in the Department of Meteorology at the University of Utah, which provides access to other Web servers around the world. Access to information in the Internet Companion is handled by the servers of the Web and, for the most part, is transparent to the client.

The client software, such as the National Center for Supercomputer Applications Mosaic software or the commercial Netscape software, has evolved rapidly during the past several years. Which client software an individual uses is primarily a matter of personal preference and convenience. To begin, an "html document," often referred to as a Web page, is accessed. After the client software is installed, some location on the World Wide Web will be defined as "Home" or the "Home Web page." This might be the Home page of a university or commercial server. Access to the Internet Companion is obtained by moving to the appropriate Web page using an option of the client software referred to as the Uniform Resource Location or simply Location: http://www.met.utah.edu/climate.html. The location of the Internet Companion Web page can then be saved as part of the client software so that it is no longer necessary to remember this cumbersome string of characters.

The Internet Companion is organized according to the structure of the book. A general index is also provided to help find particular subjects. In each chapter of the Internet Companion, text that is underlined will lead to further information. An annotated description of that information, including its type, is provided. Types range from html documents, images (in a variety of formats such as Gif or Jpeg), or short movie clips (in Quicktime, Mpeg, or Java format). Additional software may be required to access some of the more complex types of information.

A characteristic of the Web is that it is more difficult to describe than to use. So log on to your computer and try out the Internet Companion!

Index

Acid rain, 104, 106
Aerosol, 103–106
    sulfate, 52
Air-sea interactions, 55–65
Albedo, 14
    cloud effect, 24, 107–110
    ice feedback, 72–74, 76–77, 115–116
Antarctic ozone hole, 131–136
Anthropogenic forcing, 4
    fossil-fuel burning, 91–93
    deforestation, 84, 91–94
Aphelion, 77–79
Atmosphere, *see also* Climate
    chaos, 41–42
    circulation, 30–31, 128–129
    composition and origin, 50–51
    heat capacity, 58–59
    heat engine, 29–31
    interaction with ocean, 55–65
    stratification, 19
    temperature, *see* Temperature, surface

Biogeochemical cycle of carbon, *see*
        Carbon cycle
Biological pump, in carbon cycle, 86–90
Biomass burning, 84

Calcium carbonate, 88
Carbon cycle, 83–94
    biological pump, 86–90
    in present era, 91–93
    land portion, 84–85
    marine portion, 86–90
Carbon dioxide, *See also* carbon cycle
    concentration, 2–3, 70–71
    consensus view of effects of increased
        concentration, 123
    dissolved in sea water, 89–90
    emissions, 84–85
    future increases in
        concentration, 94–96
    radiative forcing due to doubled
        concentration, 101–102
    relation to surface temperature
        records, 70–71

simulations of doubling
        concentration, 118–123
Chaos, 4, 41–42
Chapman chemistry, 129–131
Chemical weathering of rocks, 91
Chemistry:
    of ozone, 129–131
    review, 144–146
Chlorine:
    and ozone hole, 133–135
    catalytic reaction with
        ozone, 130–131
Chlorofluorocarbon (CFC), 99–100,
        130–134, 137–139
Climate. *See also* Carbon dioxide; Ozone
    anomaly, 5
    change, 4
        evidence for in observed record, 6–9
        prediction of global
            warming, 113–124
        prediction of ozone
            depletion, 127–141
    effects of ENSO phenomenon, 62–65
    effects of volcanoes, 51
    example of, 4–5
    normal, 5
    models, 43–45
    paleoclimate, 67–71
    sensitivity parameter, 115–118
    simulations, 43–45, 116–125
        doubling carbon dioxide, 118–123
        paleoclimate, 72
        validation, 116–118
    variations:
        short-term, 49–65
        long-term, 67–80
Cloud:
    albedo effect, 24, 104, 107–110, 114
    condensation nuclei, 104
    effect on:
        ozone, 131–136
        radiation balance, 21–24
    greenhouse effect, 23, 104, 107–110,
        114
    imagery, 32–36

Cloud (*Continued*)
 infrared, 36
 visible, 32
 water vapor, 36
 radiative forcing, 106–110
 infrared, 107–108
 visible, 107–108
Computer applications:
 climate models, 44
 paleoclimate simulations, 72
 weather prediction, 40
Condensation nuclei, 104
Contrail, 104
Cycle:
 hydrologic, 45–47
 carbon, 83–94

Deforestation, 84, 93–94
 in carbon cycle, 91–93
Dimethyl sulfide, 105

Earth Radiation Budget Experiment, 107
Earth-atmosphere system, 15. *See also*
  Climate
 equilibrium temperature, 16
 long-term climate variations, 67–80
 short-term climate variations, 49–65
Eccentricity, 77–79
Eemian interglacial, 68
Effective radiating level (ERL), 16,
  21–24
 average, 16
 clear-sky, 22, 107
 local, 21
El Niño, 59–61
 El Niño/Southern Oscillation
  (ENSO), 62–65
Electromagnetic radiation, *see* Radiation

Feedback, 114–116
 ice-albedo, 72–74, 76–77, 115–116
 water vapor, 114–115
Forecasts:
 weather, 36–42
 climate, *see* Climate simulations
Fossil-fuel burning, 83–84
 in carbon cycle, 91–93
Freon, 99–100, 130–134, 137–139

Geochemical carbon cycle, 90–91
Glacial episode, *see* Ice Age
Global warming, 2. *See also* Carbon
  dioxide; Greenhouse
 evidence for in observed record, 6–9
 radiative forcing, 101

risks associated with, 124–125
 simulations, 118–123
Graphs, interpretation of, 144
Greenhouse:
 effect, 16, 113,
  by clouds, 23–24, 104, 107–110
  in simple model, 18
 gas, 2, 97–100
  carbon dixoide, 3, 70–71, 83–84,
   89–90, 94–96, 101–102
  chlorofluorocarbon (CFC), 99–100,
   130–134, 137–139
  methane, 98–99
  nitrous oxide, 100
  ozone, 127–141
  water vapor, 36, 45–47, 114–115

Heat capacity of oceans and
  atmosphere, 58–59
Heat engine, atmospheric, 29–31
Heat transfer, 55–58
Hole of ozone over Antarctica, 131–136
Holocene interglacial, 68
Hydrologic cycle, 45–47

Ice Age, 68, 72–80
 paleoclimate records, 73–75
Ice sheets, 72–74
Ice-albedo feedback, 72–74, 76–77,
  115–116
Inorganic nutrients, 89
Intensity of radiation, 12
Interglacial period, 68, 72
Intergovernmental Panel on Climate
  Change, 123
 projected increases of carbon
  dioxide, 94–96
Isotope, 146
 analysis, 74–75

La Niña, 61
Little Ice Age, 68
Lorenz, Edward, 4, 41

Methane, 98–99
Milankovitch theory of the ice
  ages, 75–80
Milankovitch, Milutin, 75
Mixed layer of ocean, 56
Montreal Protocol, 99, 139–140
Mt. Pinatubo, 52
Mt. Saint Helens, 52

National Centers for Environmental
  Prediction, 37

National Weather Service, 37
Net primary production, 85
Nitrogen:
    formation in atmosphere, 50–51
    in marine portion of carbon
            cycle, 89–91
    nitrate, 86–91
    nitrous oxide, 100
Nonlinear process, 42
Numbers, 143
Numerical weather prediction, *see*
            Weather prediction
Nutrients, inorganic, 89–90

Ocean:
    circulation, 56–58
    composition and origin, 50–51
    deep layer, 56–58, 87–89
    dissolved gases, 50
    heat capacity, 58–59
    marine portion of carbon cycle, 86–90
    mixed layer, 56–58, 87–89
    rising sea level, 125
    sediment cores, 74
    thermohaline circulation, 57–58,
            87–89, 125
    upwelling and downwelling, 57,
            86–90
Orbital variations, 76–80
    eccentricity, 77–79
    precession, 78–80
    tilt, 76–77
Oxygen:
    formation in atmosphere, 50–51
    in marine portion of carbon
            cycle, 86–90
    isotope analysis, 74–75
Ozone, 127–141
    as greenhouse gas, 127
    consequences of depletion, 136
    hole over Antarctica, 131–136
    natural life cycle, 128–129
    production, 129–131

Paleoclimatology, 67–75
    climate indicators, 70–72
    climate record, 67–70
    ice-age record, 73–75
Perihelion, 77–79
Photosynthesis, 146
Plankton, role in carbon cycle, 87–88
Power of radiation, 12
Precession, 78–80
Precipitation:
    acid, 104, 106

climate simulations, 118–123
    global average, 47

Radiating level. *See* Effective radiating
            level (ERL)
Radiation:
    absorbed, 14
    absorption by gases, 33–34
    albedo, 2, 14
    balance, 11–25
        and greenhouse gases, 97–100
        effect of clouds, 21
        model:
            isothermal, 16–18
            nonisothermal, 19–21
    global and annual average, 14
    infrared, 13, 36
    intensity, 12
    latitudinal distribution, 28
    power, 12
    scattering, 53
    solar constant, 14
    spectrum, 13
    ultraviolet, 13, 128, 136
    visible, 13, 32
    window, 33, 36
Radiative forcing, 100–103
    by clouds, 106–110
        infrared, 107–108
        net, 107–108
        variations with latitude, 109
        visible, 107–108
    by greenhouse gases, 101–103
    by sulfur, 105–6
Radiative transfer model, 101
Rain forests, 93–94
Rain, *see also* Precipitation
    acid, 106
Reforestation, 94
Reservoir, 45, 90–91
    carbon, 85–93
Residence time:
    carbon dioxide, 86
    equilibrium time to increased carbon
            dioxide, 92
    carbon in ocean, 91
    greenhouse gases, 98
    water in atmosphere, 45–47
    sulfur, 104
Respiration, 146

Satellite:
    cloud imagery, 32–36
    polar orbiting, 32, 107–108
    geostationary, 32

Scientific notation, 143
Sea surface temperature, 59–61. *See also*
    Ocean; Temperature, surface
Sea level, rising, 125. *See also* Ocean
Sediment core, 74
Sensitivity parameter, climate, 115–118
Simulation:
    anomaly, 119
    control, 119
    of doubling carbon dioxide, 118–123
    with radiative transfer models, 101
Skin cancer, effects of ozone
    depletion, 136
Solar constant, 14, 78
    variations, 68
Stefan-Boltzmann Law, 16–17, 101, 115
Stratosphere, 1, 127–141
    circulation, 31, 128–129
Sulfur, 103–106
    emission sources, 104–106
    radiative forcing, 105–106
    sulfate aerosol, 52, 103–106
    sulfuric acid, 52
Sunset, effects of volcanic eruptions, 52
Sunspots, 68

Temperature:
    equilibrium, 16
    ground, 17
    paleoclimate indicators, 70–75
    surface, 2
        climate simulations of, 118–123
        global record, 6–9, 67–70, 73–75
        in climate models, 116–118
        trend, 7

    uncertainties, 8
    variations:
        short-term, 49–65
        long-term, 67–80
    vertical structure, 6
Thermohaline circulation, 57–58, 87,
    125
Tilt, 76–77
Troposphere, 1
    circulation, 30

Ultraviolet radiation, *see* Radiation,
    ultraviolet
United Nations Environmental
    Programme, 137
Units, 143–144

Volcanoes, 49–55
    effects on climate, 51–55
    major eruptions, 51–52
    ozone depletion in stratosphere, 135
    role in composition and origin of
        oceans and atmosphere, 50–51

Water cycle, 46
Water vapor
    as greenhouse gas, 36, 114–115
    residence time in atmosphere, 45–47,
        86
Weather prediction, 36–42
Weather, 27, 37–42
Weathering, chemical, 91
Wind, 27–31
    effect on ocean currents, 57
World Meteorological Organization, 37